Design Science and Innovation

Series Editor

Amaresh Chakrabarti, Centre for Product Design and Manufacturing, Indian Institute of Science, Bangalore, India

The book series is intended to provide a platform for disseminating knowledge in all areas of design science and innovation, and is intended for all stakeholders in design and innovation, e.g. educators, researchers, practitioners, policy makers and students of design and innovation. With leading international experts as members of its editorial board, the series aims to disseminate knowledge that combines academic rigour and practical relevance in this area of crucial importance to the society.

Prabir Mukhopadhyay

Human Factors in Tourism: A Layperson's Approach

 Springer

Prabir Mukhopadhyay
Design Discipline
PDPM Indian Institute of Information
Technology Design and Manufacturing
Jabalpur, Madhya Pradesh, India

ISSN 2509-5986 ISSN 2509-5994 (electronic)
Design Science and Innovation
ISBN 978-981-99-7065-0 ISBN 978-981-99-7066-7 (eBook)
https://doi.org/10.1007/978-981-99-7066-7

© The Editor(s) (if applicable) and The Author(s), under exclusive license to Springer Nature Singapore Pte Ltd. 2024

This work is subject to copyright. All rights are solely and exclusively licensed by the Publisher, whether the whole or part of the material is concerned, specifically the rights of translation, reprinting, reuse of illustrations, recitation, broadcasting, reproduction on microfilms or in any other physical way, and transmission or information storage and retrieval, electronic adaptation, computer software, or by similar or dissimilar methodology now known or hereafter developed.
The use of general descriptive names, registered names, trademarks, service marks, etc. in this publication does not imply, even in the absence of a specific statement, that such names are exempt from the relevant protective laws and regulations and therefore free for general use.
The publisher, the authors, and the editors are safe to assume that the advice and information in this book are believed to be true and accurate at the date of publication. Neither the publisher nor the authors or the editors give a warranty, expressed or implied, with respect to the material contained herein or for any errors or omissions that may have been made. The publisher remains neutral with regard to jurisdictional claims in published maps and institutional affiliations.

This Springer imprint is published by the registered company Springer Nature Singapore Pte Ltd.
The registered company address is: 152 Beach Road, #21-01/04 Gateway East, Singapore 189721, Singapore

Paper in this product is recyclable.

To the Lotus Feet of Shri Ramakrishna Paramhanhsa Dev, Pronam To You.
Late Dhirendranath Mukhopadhyay (My Father),
and
Mrs. Meena Mukhopadhyay (Mother), Dada, Mamoni, Duggi and Leto.

Preface

In the tourism and hospitality industry, the focus is at on people whom we call tourists, guests, travelers and many names. Till date, there are not many books which talk about the human factors issues in tourism, with specific focus on how to enhance customer satisfactions. A few publications that are available exclusively focus on the human factors of injury in this sector with specific directions for preventing them. There is a distinct dearth of reference materials which exclusively talks about the application of human factors in tourism and that too in a language which is easy to understand and apply. This book is an attempt in that direction. It's written in an easy to understand language. The book can be used by anyone from any background. The readers need not have any prior knowledge of the subject to read and understand the different chapters. Each chapter is written in a storytelling format with images which have been used for representation purpose only and for building the storyline. There are only six chapters, and the book is not bulky and should not make the readers jittery in going through it. This would be useful for all those associated with the tourism and hospitality industry. It's not a replacement for the textbooks on human factors available in the market. The purpose of this book is to introduce the subject to those who have never studied this before in the tourism and hospitality industry. This can be used as a reference book at their work place for augmenting customer satisfaction and increasing efficiency. The book focuses more on the intangible issues of human factors and its application rather than only on injury and its prevention. This is where the book is different from any other books in the market.

The writing is in a story telling format, with examples from real world, which should make the book interesting to read and at the same time easy to apply in the real world. Each chapter touches upon different aspects of tourism. They start with an "overview" section which prepares the readers with what they can expect out of the chapter and ends with "key points" which are a gist of the chapter. This is followed by practice questions with ergonomic directions which would help the readers test out how much they have understood. The last chapter is dedicated to different type of exercises with human factors directions. The exercises have been carefully selected from different aspects of tourism and its human factors application.

All the images in this book have been taken from royalty-free sources. They are for representation purpose only and builds the storyline.

I hope the readers enjoy reading the book. Any comments and suggestions are always welcome.

Jabalpur, India
Dr. Prabir Mukhopadhyay
prabir@iiitdmj.ac.in

Acknowledgements The line diagrams have been drawn using Google Auto Draw free online AI tool. We sincerely acknowledge Google Auto Draw. All the images used in this book have been taken from royalty-free sources. We sincerely acknowledge all the photographers for allowing us to use their images for free.

The feedback given by all anonymous reviewers is also deeply acknowledged.

Contents

1 Your Users Your God 1
   1.1   The Canvas of Human Factors in Tourism 1
   1.2   What Users Know 3
   1.3   Challenges of Drawing Customers and Retaining Them 4
   1.4   The Weak Link in Tourism and Hospitality 5
   1.5   Areas Where Human Factors Can Intervene 6
   1.6   The Human Factors of Tourism and Hospitality 6
   1.7   What Users Want 7
   1.8   The Tourism Types and Users' Needs 8
   1.9   Tourism Customized and Group 10
   1.10 Tourism the Journey from Planning to Return and the Touch Points 10
   1.11 Travel Modality 13
   1.12 Food, Photo Shoot, Explanation and Experience 14
   1.13 For Some It's the Destination for Some it's the Journey 15
   1.14 Safety Aspects in Tourism 15
   1.15 Snowball Effect in Tourism One Tour Ends and Planning for Another Begins 16
   1.16 The Problems with Excess Information: We Chew More Than We Can Digest 16
   1.17 Key Points 16
   1.18 Practice Session 18
   Bibliography 19

2 First Love with Accommodation 21
   2.1   Accommodation: The Humane Touch: Lobby, Counters, Clocks 21
   2.2   Locating the Place in the Actual Context 24
   2.3   The First Interaction Wow 25
   2.4   Staffs Look and Feel 27
   2.5   Importance of Check-In and Check-Out Counter 30
   2.6   How to Ensure Proper Crowd Formation 30
   2.7   Attention Span of Guests and How to Deal with It 32

ix

2.8	The First Impression of the Room	33
2.9	Elements that Makes the Room Beautiful	33
2.10	Humans and Room Ventilation and Illumination	35
2.11	The Bed	35
2.12	The Table and Chairs	36
2.13	The Bathroom and Its Elements	37
2.14	Bathtub	37
2.15	Remote for TV and AC	37
2.16	Window Curtain	37
2.17	Light Control from the Bed	38
2.18	The Bathroom Direction at Night	38
2.19	Replenishing Tea, Coffee, Toiletry, Towels and Soaps	38
2.20	Means of Detecting Missing Elements in Room	38
2.21	Key Points	39
2.22	Practice Session	39
	Bibliography	41

3 Host and Hospitality 43
3.1	Fitting Your Guests and Staffs	43
3.2	Injury of Staffs and Prevention	45
3.3	The Dining Room	45
3.4	Elements of Food Design and Layout	47
3.5	Equipment Handling	48
3.6	The Buffet Chaos	48
3.7	Crowd Flow Management	49
3.8	Special Occasion in Dining Area	50
3.9	Replenishing Fast Depleting Foods	50
3.10	The Food Design Elements	50
3.11	Ambience Creation	50
3.12	Key Points	51
3.13	Practice Session	51
	Bibliography	52

4 Food and Tour 53
4.1	Kitchen and Dimensional Issues	53
4.2	Manual Material Handling in Kitchen	55
4.3	The Correct Human Factors of Serving Food and Beverages	56
4.4	Cleaning Activities and Postural Issues	57
4.5	Maintaining Tone of Muscles	57
4.6	Slips Trips and Fall and Their Prevention	58
4.7	Work–Rest Cycle	58
4.8	What Defines Enjoyment?	58
4.9	Dealing with Group and Individual	59
4.10	Creating the Background for Travel and Tourism	60
4.11	Local Cuisine While on Tour	61
4.12	Change in Travel Plans Due to Natural Calamity	61

	4.13	Information in Space About Tourist Site	61
	4.14	Elderly and Challenged Tourists	62
	4.15	Different Types of Sites, Heritage, Natural, Performance and Mementos to Remember	62
	4.16	Tips for Those Who Provide Services	62
	4.17	Key Points	63
	4.18	Practice Session	63
	Bibliography		65
5	**Reducing Travel Uncertainty**		67
	5.1	The Uncertainty in Travel	67
	5.2	The Service Angle and Human Factors in Hospitality	69
	5.3	What Customers Aspire?	70
	5.4	Peoples Weird Choice and Demands	70
	5.5	Why People Prefer Some Over the Others	70
	5.6	That Something Extra for Your Customers: Customer's Sweet Spot	71
	5.7	The Humane Touch and Winning Hearts	71
	5.8	Information for Reducing Uncertainty	72
	5.9	Global Tourist	73
	5.10	Cultural Angle	73
	5.11	Key Points	73
	5.12	Exercises	74
	Bibliography		74
6	**Exercises in Human Factors**		75

About the Author

Dr. Prabir Mukhopadhyay He holds a Ph.D. in Industrial Ergonomics from the University of Limerick, Ireland. He worked at the National Institute of Design Ahmedabad as a faculty in Ergonomics for two years before leaving for his doctoral study at the University of Limerick, Ireland. On his return from Ireland, he joined as Faculty at the Post Graduate campus of the National Institute of Design at Gandhinagar. After working there for around five years, he joined Indian Institute of Information Technology Design and Manufacturing Jabalpur as Faculty in Design Discipline. He has successfully completed several research and consultancy projects, and some of his clients are the Indian Railways, Self Employed Women's Association, and the United Nation Industrial Development Organization, etc. He has authored five single authored books till date: *Ergonomics for the Layman: Application in Design*, *Ergonomics Principles in Design: An Illustrated Fundamental Approach*, *Ergonomics in Fashion Design: A Laypersons Approach*, *Visual Ergonomics for Communication Design: A Laypersons Approach*, and *Application of Ergonomics in Handicraft: A Laypersons Approach*. He may be reached at prabirdr@gmail.com

List of Figures

Fig. 1.1	Pyramidal structures of human factors in tourism depicting different components	6
Fig. 1.2	User's needs and wants in tourism is key to success. Photo by Parv Choudhary: https://www.pexels.com/photo/people-riding-elephant-2870167/	7
Fig. 1.3	Different mental models of tourists. Photo by Aleksandar Pasaric: https://www.pexels.com/photo/people-walking-on-the-street-between-buildings-2385210/F	9
Fig. 1.4	Luggage divided into quadrant and the visual of the same presented to clients for easy retrieval. Photo by Kindel Media: https://www.pexels.com/photo/blue-luggage-with-folded-clothes-8212229/	11
Fig. 1.5	Modular luggage system integrated with main luggage. Photo by Anastasia Shuraeva: https://www.pexels.com/photo/a-suitcases-with-gold-handles-on-a-wooden-floor-5705068/	12
Fig. 1.6	Travel thread from home and back to home. Photo by Tatiana Syrikova: https://www.pexels.com/photo/retro-globe-with-handmade-toys-against-gray-background-in-room-3932932/	14
Fig. 1.7	One tour experience leading to another tour. Photo by Malte Luk: https://www.pexels.com/photo/man-in-brown-top-beside-railroad-2432299/	17
Fig. 2.1	Reception staff at hotel with a smile is a welcome gesture. Photo by Cedric Fauntleroy: https://www.pexels.com/photo/a-woman-talking-to-a-receptionist-4266932/	22
Fig. 2.2	Direct eye-to-eye contact at the reception plays a very important role. Photo by Mikhail Nilov: https://www.pexels.com/photo/a-female-tourist-talking-to-the-female-receptionist-of-the-hotel-7820311/	23
Fig. 2.3	Guest standing at the reception counter. Photo by Mikhail Nilov: https://www.pexels.com/photo/a-guest-talking-to-a-woman-at-the-reception-area-7820376/Figure	25

Fig. 2.4	Tourists trying to locate accommodation through signage system on the road. Photo by Davis Sánchez: https://www.pexels.com/photo/woman-standing-beside-alto-signage-2006953/	26
Fig. 2.5	First interaction and the customer's wow! Photo by RODNAE Productions: https://www.pexels.com/photo/a-female-front-desk-receptionist-smiling-to-a-client-7755676/	27
Fig. 2.6	Luggage trolleys for carrying heavy luggage. Photo by Mikhail Nilov: https://www.pexels.com/photo/a-woman-talking-picture-at-the-reception-area-7820313/	28
Fig. 2.7	Well-groomed staff generates a better image of the place. Photo by Ketut Subiyanto: https://www.pexels.com/photo/woman-in-gray-t-shirt-checking-in-and-smiling-4907441/. Extra Counter with concavity: Photo by Pavel Danilyuk: https://www.pexels.com/photo/men-buying-movie-tickets-7234477/	29
Fig. 2.8	Crowd control through design of reception counter. Photo by Pavel Danilyuk: https://www.pexels.com/photo/men-buying-movie-tickets-7234477/	31
Fig. 2.9	Crowd control through design at a counter. Photo by Lachlan Ross: https://www.pexels.com/photo/people-gathering-near-counter-at-festival-6510339/	32
Fig. 2.10	Room as the guests visualize. Photo by Pixabay: https://www.pexels.com/photo/bedroom-door-entrance-guest-room-271639/	34
Fig. 2.11	Guests looking at the bathroom mirror. Photo by Andrea Piacquadio: https://www.pexels.com/photo/woman-in-white-tank-top-while-looking-herself-at-a-mirror-3811809/	34
Fig. 2.12	Bed making and dimensional requirement considering movements. Photo by Liliana Drew: https://www.pexels.com/photo/housekeepers-changing-sheets-in-hotel-room-9462740/	36
Fig. 3.1	Adequate space between people while dining. Photo by fauxels: https://www.pexels.com/photo/sharing-cherry-tomatoes-3184188/	46
Fig. 3.2	Linear movement of staffs without crossing. Photo by energepic.com: https://www.pexels.com/photo/woman-holding-tray-of-wines-2991157/	49
Fig. 4.1	Heavy work of grinding seeds being performed little below the elbow height. Photo by Uriel Mont: https://www.pexels.com/photo/crop-faceless-man-pouring-water-into-coffee-grinder-6271318/	54

Fig. 4.2	Light work of vegetable chopping being performed around the elbow. Photo by Sarah Chai: https://www.pexels.com/photo/faceless-woman-chopping-bell-pepper-during-lunch-preparation-in-kitchen-7262911/	55
Fig. 4.3	Precision work of cake decoration being performed around the chest region. Photo by Tim Douglas: https://www.pexels.com/photo/young-woman-putting-decoration-on-top-of-homemade-cake-6210907/	56
Fig. 4.4	Elbow-to-elbow distance roughly defines one person working space requirement. Photo by Elle Hughes: https://www.pexels.com/photo/man-and-woman-wearing-black-and-white-striped-aprons-cooking-2696064/	57
Fig. 5.1	Uncertainty of travelers decreases the pleasure of the tour. Image by **Masashi Wakui** (https://pixabay.com/users/masashiwakui-4385858/?utm_source=link-attribution&utm_medium=referral&utm_campaign=image&utm_content=2014617) from Pixabay (https://pixabay.com/?utm_source=link-attribution&utmmedium=referral&utm_campaign=image&utm_content=2014617) https://pixabay.com/photos/japan-osaka-night-asia-landmark-2014617/	68
Fig. 5.2	Providing service to customers before they ask for it. Image by: **cottonbro studio** (https://www.pexels.com/@cottonbro/) https://www.pexels.com/photo/server-pouring-hot-drink-6466292/	69
Fig. 5.3	Guests enjoying food over a music which sets the tone of the ambience. **Negative space** (https://www.pexels.com/@negativespace/) https://www.pexels.com/photo/glass-architecture-windows-lines-34112/	72

Your Users Your God

1

Overview

This chapter gives the readers an overview of the subject human factors and its application in different areas of tourism and hospitality sector. Human factors as a subject can not only help identify the areas where humanization is possible both in terms of the tangible and the intangible, but it can also indicate the point of vulnerability in the entire area where ergonomic intervention could prove to be beneficial. The users, who might be our customers or our staffs involved in these sectors, are the prime stakeholders. Thus, an insight into their needs and wants and how one could cater to that are best explained in this chapter. A broad overview of the ergonomic issues in tourism, involving the planning and the journey phase, is discussed. It discusses the ergonomic issues in customized tourism based upon user's needs. The chapter takes the readers through different issues in tourism like safety and other adventure that tourists would like to venture into. The different aspects of tourism including food, photo shoot, from an ergonomic perspective are explained here. The chapter also talks about the issues of visual human factors of providing optimal information to the tourists so that they are able to optimize their time and enjoy to the fullest.

1.1 The Canvas of Human Factors in Tourism

People have a concept nowadays that they prefer working hard and partying hard. It's not that everyone party, but they would like to relax. Thus, human relaxation has different forms, and not every user prefers the same way of relaxation. Then are another set of users we call them "service providers" who provide their customers with what they need. A seamless integration between the customers and service providers leads to better experience by customers when it comes to tourism and hospitality. There are two aspects; one is travel, and the second is accommodation

© The Author(s), under exclusive license to Springer Nature Singapore Pte Ltd. 2024
P. Mukhopadhyay, *Human Factors in Tourism: A Layperson's Approach*, Design Science and Innovation, https://doi.org/10.1007/978-981-99-7066-7_1

and food. These two attributes are important for people when they step out if their homes, because they are doing this leaving behind the comfort and convenience of their homes. As people have a tendency of expecting whatever they have in their familiar surrounding even while traveling or staying somewhere else, it becomes difficult to satisfy them. This is a big challenge in the hospitality industry. People want same thing what they get at home and on the road but at the same time when it comes to tours and travels, they also look for newness and variation. Their attention span becomes very short, and they tend to get easily frustrated if things do not meet their expectations. As every individual is different the situation becomes all the more complicated. This is where the subject human factors, which is all about the interaction between people, their needs and wants in a specific context, become very important for the sector.

If I ask you what is the yardstick which measures a good hospitality, or an enjoyable or memorable journey? It is difficult but has to be done if the industry has to survive. In these cases, we take feedback from users as to what was good and what went wrong and then try to improve upon them. But if we could have had a template to follow which would ensure maximum user satisfaction in tour and hospitality would that be better? That would help the industry prepare much better and not leave any room for disaster and bad reputation? Human factors aims at that. It looks at the tourism and hospitality sector from the view point of users, what they would like or dislike, their preference, way of working, expectations, etc. and then give the industry some concrete directions how to go about. All these are done with an eye to enhance service and customer satisfaction. In this sector your customer is always your God. Human factors is all about the "user" your God!!

People want information in advance before they venture out. How is the place, cost implications, places to see, local cuisine, safety issues, weather, clothing to be carried and host of similar information. This information needs to be presented to the users before they embark upon their journey as that gives them a sort of "picture" (called mental model) in front of their eyes as to what they would visualize and experience in the place. This is where human factors can play a major role in designing and layout of the tourist place or the promotional materials or the office in converting that into a real world where people can see them. Thus colors, people, photographs of cuisines and other elements which represent the place are normally used to design the place. Mr. Leto is visiting a hill station name Orang. It had small hills with a large number of waterfalls and loads of coconut trees. The place is also inhabited by a large number of peacocks. Now to give a feel of the place all materials should contain at least some of these elements that is coconut, peacock feathers, waterfalls because these are the identity of the place and hence can act as the trigger or framework which would elicit that feeling in your users as to what they are likely to get when they in reality go to the place. If possible, tourist offices can use audio visual projections for showcasing these elements in the space. So, human factors keeping the user at the back of the mind tries to play with different exciting elements that would excite people. Similarly for giving people a feel of the hotel the elements inside the hotel like gym, swimming pool and the furniture pieces inside the rooms, the bathrooms, etc. could go a long way in alluring people as to what is the level of luxury or comfort the hotels can provide. These help the users take prudent decision in where to go, how to go, where to stay, etc. People see these and start fanaticizing themselves in the midst of these. After all the body needs some level of rest after a hard and monotonous work for months? This helps the person to come back to work rejuvenated and with high spirits. Thus, while we talk about tourism and hospitality one needs to be careful about people's needs and wants and what is the information they want and we should be very precise on those aspects of it, neither more nor less. When looking at this bigger picture or canvas of tourism and hospitality we find that there are travelers, staffs, places, cuisines, cultures, safety, hospitality, how travelers are accepted by the locals and the weather. If looked at carefully the canvas of this sector consists of different types of users, and this is where human factors can help in a seamless integration of variety of people and the elements in the canvas to make the canvas robust and less vulnerable to error. The problem in this sector is that once your customer faces a problem it becomes a bad mark in his memory and she/he would avoid it for even and the industry loses a customer forever.

The canvas of the sector then needs intervention from the user's perspective at many different levels. First layer you need to know the user's needs, wants, capacity, limitation, budget and other needs. In the second layer your focus should go on the variety of users, male, female, elderly, local foreign, tourists with medical conditions, etc. The third layer your focus should shift to what experience the tourists expect from you and how it could be addressed in the best possible manner. This is where tourists need facility and experiences that they are used to getting at their home, which includes the cuisine but the same time expect some newness as well. In the fourth layer of this canvas your focus should go on the experiential part or pleasure part of the sector. This is the most important part why tourist travels to far off places and unwind from their busy work schedule. So essentially human factors helps you to look at all the four layers and address each with equal emphasis, in order to be successful in giving the best possible services to your users.

1.2 What Users Know

The problem or the good part is that tourists have some information with them before they start their journey. This is natural as users when they embark on any journey outside their routine ones, they are faced with lots of uncertainties and thus they look for information which would reduce their uncertainty. This is where the problem emanates that if users have more information, then they tend to get lost in this sea of information, not knowing what is useful and what not. If information is less than the requirement then, it does not reduce their uncertainty. Thus, we need to know what our users already know and then provide the requisite information in the format that they would be able to use it with prudence.

One of the many ways of knowing what the users already know is to survey the target users and get an insight into this. This lets us know the way users are thinking, their pattern of utilizing the information, what more they expect, information related to which specific aspect they expect and so on. Mr. and Mrs. Goody are planning for a vacation to a beautiful place Sea Saw which is a sea beach. They are elderly couple, and Mr. Goody is aged 82 years and Mrs. Goody 78 years. Both of them are a little jittery when it comes to travel but at the same time, they are bored staying at home and feel that they should go for a vacation. This is where human factors should come to the rescue. Your approach through human factors would be to understand the user's pain point and then provide them with the information that reduces their uncertainty. If you mention to them that at every place there are facilities for elderly access, there are escorts to help elderly people move around with their luggage they would be much relieved. Moreover, a bit of information on medical facilities available during the journey as well as at the tourist spot would act as s moral boost for this type of tourist. On the other hand, Mr. and Mrs. Dare are in their early 30. They work in the same office and are yet to marry. They live a very hectic life and decide to take a break for a Jungle Safari to a jungle named Green Belt. These people are looking for adventure and to them uncertainties in the journey is what they would like to cherish. So, travel and tourism for them should be with less of information and more emphasis on the information which only talks about new ness, new adventure, new food or in other words uncertainty. So, your user needs are varied. Such users would even aspire for new adventure, cuisine; places, etc. because they do not have any health restrictions and would only explore more and more newness. This newness to them gives them the pleasure. Whereas for the elderly who have explored the world, would like to explore it at a very slow pace, at their own slow pace and without much of a travel but confining themselves to a particular location or a few locations only.

1.3 Challenges of Drawing Customers and Retaining Them

There is a cut throat competition in the tourism and hospitality market. There are company which are providing similar facilities in terms of travel, accommodation, cuisines, adventure, etc. at similar price tag. So how would the customer decide with whom to go? This is where those softer sides of the customers have to be taken into account. So, what does a user want when they tour? There are two aspects of any users. The physical aspects which deal with their level of fitness, disability if any, dimensions (overweight) which might come in the way of travel or accommodation. Users are concerned about this at times. If you add to these physically challenged users like those who are mobility impaired, sensory impaired (blind, deaf dumb, etc.) then the presentation of information should be in the form of different modality. For example, for the blind information to be given through auditory route, for deaf and dumb through visual route and so on. The combination of the physical and mental attributes in humans comprises of the total user or customer base. Now if you have to draw your customers toward you, then you have to go and provide that extra mile for them. If your travel and tour can provide customized solution for the wheel chair bound to get a feel of the sea water, or if your company can give the bling a look and feel of the snow, that's something unique that could be offered. This unique comes in the form of offering some customized services that match their needs. These special or customized services remain in their mind and create an image of the company for a long time. This image is transferred to other customers and your customer base increases. In this industry you have to offer your customer something very close to their heart which others don't. That's how you draw your customers toward you. This is not easy but a careful interplay of information (curated), planning, personal touch and interaction and follow up all through the journey and even when the journey ends. For example, if you ask your clients at the end of the tour when they reach home Sir have you reached safely, how was your experience with us? This type of personal touch goes a long way in creating the long-lasting impression in the minds of the people.

What about those who are always complaining, never happy, have problems with everything? You need to treat them with care and love. Remember all humans get irritated when there is lack of feedback to them. If they are looking for something as they are unable to find it, they get irritated. You need to establish this feedback system with them, and this actually gives you a cutting edge over the others when it comes to people selecting your services for tours and travel over the others. That means you have to be different from what others do, and this could be achieved through constant feedback from your users.

1.4 The Weak Link in Tourism and Hospitality

The place where we all make a mistake is when we start blaming users for every odd thing. If customers are not happy, misbehaving, irritated we just put the onus on them saying that they are bad people. Always remember that humans do not deliberately behave in a particular bad way. They do so when they lack feedback in the entire systems loop of tours and tourism. The systems perspective is just like a triangle where we have customer at the apex, tour operators at one corner and the tangible elements of places, objects, etc. on another corner. All concentration should be at the sharp end but focus should equally go at the end where tour operators decide upon and chalk out the plans and programs for the customers. Mistakes or dissatisfaction of customers happens because they do not get timely feedback from the staffs and staffs do not act in accordance with this feedback of customers. The weakest link in the industry is because of the wrong perception that all customers are the same, which in essence is very wrong. If you perceive that everyone is same and the same services and its quality would make everyone happy, this becomes the most vulnerable point in the industry. Customization is the key in the industry to make the weak link of human variation robust and thus lead the sector forward in leaps and bounds.

1.5 Areas Where Human Factors Can Intervene

Human factors in physical terms can guide you toward dimensional aspects of products, space, keeping in mind movement pattern and usage pattern of users. The subject can indicate where and how one needs to be extra cautious for elderly, challenged group. How is information to be presented, how much and in what format so that people can easily understand? When is a running commentary better and when should one keep silent during the tour so that users can enjoy and get absorbed with the surroundings. Where should users be left on their own and where should hand holding of users start.

1.6 The Human Factors of Tourism and Hospitality

Human factors tend to make the weak link in the system robust. It talks about personal grooming of staffs, dimensional and force attributes of products and space, manual material handling in food and beverage units, movement of people, group formation and crowd interaction. It talks about changing the mood of a gathering, creating a multisensory ambience that people remember, and how to ensure customers come back time and again to enjoy and savor the same moments.

The bigger picture that human factors lets you all allow is with specific focus on the customers and also all the people associated. It can be perceived like a pyramidal structure (Fig. 1.1). There is a context which could be the transportation modality like bus, car, train, etc., the tour in itself where people move from place to enjoy, accommodation or shelter where people stay. There are people associated and they are the tourists/customers and the staffs who provide these different services. Thus, the structure looks like a triangle with a sharp end and a base. So, at the sharp end are your customers (tourists), one corner is the modality of travel, the other corner your shelter. More or less, these comprise tourism. At the base of this pyramid are the people in the hospitality industry who take decisions, on what services to offer to clients. The service providers are both at the base and at the sharp end of the pyramid providing services to the client. This is the bigger picture which human factors actually is. Thus, your client's satisfaction, needs and wants are all interlinked with these different elements and not on any one factor.

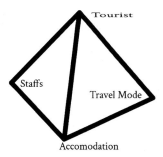

Fig. 1.1 Pyramidal structures of human factors in tourism depicting different components

1.7 What Users Want

Thus, the key to success is to know what your users want (Fig. 1.2). You have to know their characteristics and have to deal with them alike. People while on tours and travels have a different mood and mentality which needs to be respected. At times too much happiness and excitement can create chaos and discomfort for others which need to be sorted out. This is where tour operators need training in dealing with the situation in the backdrop of their knowledge and experience in the relevant field. At times getting to know the users wants and needs one has to enter into their minds and their hearts. The softer side of human factors helps you in that pursuit. For example, always tell your customers that they are right. This approach solves many customer grievances.

Fig. 1.2 User's needs and wants in tourism is key to success. Photo by Parv Choudhary: https://www.pexels.com/photo/people-riding-elephant-2870167/

1.8 The Tourism Types and Users' Needs

There are different types of tourism and each is meant for different tourists. For example, there are foreign and local tourists and the needs and wants of these tourists are different depending on their culture and stereotype they come from. There is in general tourism in general where people visit different places of interest depending on historical values, jungle safari, adventure tourism, etc. At times people prefer combining all different tourisms into one unified package so as to save on time and money. A new tourism has now emerged known as medical tourism where users are guided toward cost-effective medical treatment along with all information related to accommodation, food, etc. for the persons accompanying the patient. At times this is also clubbed with local tourism once the patient recovers, and thus one is connected to main stream tourism in this way. Your customers also ways expect what they get and what they are used to getting. So, if you able to provide them something extra, that's an extension to their expectation they cherish that. If you provide them something extra that is not an extension of their expectation then they feel disappointed. So, a careful trade-off between surprises and expectations should be accommodated so that users enjoy what s provided to them.

So, what should be the approach in tourism from the user's perspective? Users always have or prefer to rely on mental models (image of a place, journey, experience, excitement, etc.). In tourism you have to bank upon this existing image or model (Fig. 1.3) and then try to build upon the same. There could be two set of users. The first one would have a fixed plan and mental model of what they would like to see, where they would like to go. This happened because either they were briefed by their friends or relatives or the experiences narrated to them have been etched in their minds. This happens when you are repeatedly narrated some incidences over a period of time. This model is difficult to break completely because it gets the advantage of time along with them. But the good part is that as a professional you can bank upon this existing model or image already etched in their brain with some extra images. For example, if they have been briefed about a hill station you can show them pictures, narrate them stories about adventure tourism available at the same place. You might as well tell them about some exotic cuisines or music of the place which were not briefed to them. In this way your marketing of the unknown augmented by what they already know helps them to build a better mental model of the place they would like to venture. Taking off from here you might also expand your marketing further by giving them some mental model of similar places but with different levels of newness and excitement. When it comes to medical tourism, the plan is different. Users are tensed, and looking for best treatment plan and cure. If precise information is provided this is easy to process for the target users under stress and they appreciate that aspect. Sometimes further advice could be given to the patient party that after they are cured and declared medically fit what are the nearby places, they could visit either for sign seeing or for shopping or for religious purposes. A complete plan could be chalked and presented to the client so that they have the plan in front of them and based on how

1.8 The Tourism Types and Users' Needs

the treatment progresses they can take a prudent decision accordingly. Decision making by users becomes easy when users can see the options available in front in an easy-to-understand manner. This is where the user's needs and the tourist planning play an important role.

Once the journey starts users expect things to happen as per their expectations. In this stage the information needs to be presented to the user in steps and stages as they move on. A proper feedback and reconfirmation to the tourist either through visual, audio or tourist guides helps them. The information system in the space

Fig. 1.3 Different mental models of tourists. Photo by Aleksandar Pasaric: https://www.pexels.com/photo/people-walking-on-the-street-between-buildings-2385210/F

if properly aligned at regular interval could give effective feedback to the tourist. When such people move in space either on foot or in vehicles they are always on the lookout for information like where are I, where is the target, how long from here, anything else apart from this and so on?

1.9 Tourism Customized and Group

Every human wants to be treated in a special manner and in tourism this is much more prominent. Solo travelers can be helped with tourist plans by providing them with information in the form of pictorials, with very brief descriptions. These tourists would prefer specific guides along with them who would help them with the history of the place and also help in venturing into the unknown. At times the guide could also act as a friend for patient hearing of the tourist when the tourist feels very lonely. The guide can tell the story or history or incidences of different places which the tourists can relate and thus enjoy it better as they are able to recall it much better. For example, if the tourist is taken through a mountain and they are told about a film sequence which was shot there, the tourist can immediately connect that with what they see. Later on, when they go back from the tour, they are able to recollect the place much better in terms of retrieving the information with reference to the movie sequence.

In group tourism one has to be careful that likeminded people are in the group. Too much of group diversity can disturb the group dynamics in tour and can lead to problems. Thus, the tour operator has to ensure that tourists with similar interests and mental models are part of the team else the group dynamics breaks and the tours becomes a complete chaos. In case varied user groups are to be accommodated, it's better to chalk out a fixed itinerary so as to ensure that everyone follows the same and deviations should not be permitted. This would ensure that the tour operates smoothly without any glitches.

1.10 Tourism the Journey from Planning to Return and the Touch Points

Your users want a seamless tour and would not expect any glitches during the tour. They would like to have everything in proper format as was promised to them, or as was visualize by then before they embarked upon the journey. The first step is packing your bags. You can help your clients in packing the essentials so that they travel light but carry all that is needed. This optimization could be facilitated by "itemizing" and "categorizing" the belongings. First let us take "itemization". This means deciding and making a list of those items as per their priority that should go into your luggage. For example, in order of priority it could be Travel documents, money, cards, medicines, spectacles, self-monitoring kits (e.g., blood glucose meter, if one is diabetic). The garments could be further itemized as undergarments, night garments, outer garments, woolens, caps,

1.10 Tourism the Journey from Planning to Return and the Touch Points

mufflers, hand gloves, socks, shoes, slippers, hankies, etc. This list could be customized depending on individual needs. Once the list is done then the items could be categorized further in terms of external and internal clothing, accessories, medical and emergency, finance, etc. The advantage of making such a category while packing is that retrieval of each item becomes easy when the luggage is opened. So, you can make a nice list for your customers and hand it over to them. The list could be further schematically divided into different quadrants indicating the items packed within each quadrant so that emergency retrieval of any item is easy (Fig. 1.4).

The next stage in tour planning is to ensure that the customer has light baggage which is easy to carry. Here the rule of the thumb is that it's better to carry separate small luggage rather than carrying one heavy luggage which is difficult to handle and might even lead to injury. Our suggestion is luggage's on wheels are always better, provided the same need not be dragged along the staircases. It's better to go for a modular approach. For example, a larger luggage could have three small attachments with it. These three attachments can detach and function as independent luggage as and when required. This facilitates easy modularity as and when required. For example, when tourists go for shopping, they are unaware for the new load of luggage they are accumulating. Thus, after their shopping is over

Fig. 1.4 Luggage divided into quadrant and the visual of the same presented to clients for easy retrieval. Photo by Kindel Media: https://www.pexels.com/photo/blue-luggage-with-folded-clothes-8212229/

and when it comes to packing their luggage, they suddenly realize that either they have no extra space to accommodate the items or the existing luggage has become very heavy to handle. A modular approach as mentioned can absorb some of the extra items and function as independent unit (Fig. 1.5).

Once the journey starts the luggage and the user becomes an integral part. So now the design of the luggage should ensure that the user gets constant feedback of the luggage and does not loose sign of it. One can give the luggage some odd features like special fluorescent colors (enhancing visibility from a distance); some special graphics, giving the luggage a different form or attaching some protruding object makes the luggage stand apart from the crowd especially when users have to collect it from conveyer belts. This is the stage when the luggage and the tourist become an integral part and are perceived as one. For example, tourists carry their backpack which houses their camera and other kits essential for travelers as they need to access them regularly. Such bags should be designed such that once the hand is inserted the mere feel of the hand should be able to locate the required items inside the back. Thus, care should be taken to compartmentalize the bags and

Fig. 1.5 Modular luggage system integrated with main luggage. Photo by Anastasia Shuraeva: https://www.pexels.com/photo/a-suitcases-with-gold-handles-on-a-wooden-floor-5705068/

ensure no sharp edges which might cut or bruise the fingers. Each object should be held in the respective quadrants so that while pulling out specific items the others are not dislodged from their position. This is the stage where the users actively participate in the tour and tries to reap the maximum out of all the tour has to offer. As the tour comes to the end the users become desperate to see and record more just before it come to the end. This is where the users need handholding and customized tour planning needs to be suggested to them, so that based on their interest areas they could select what they want and reject what they don't. So, your job as a tour operator is to provide a bird's eye view of the tour and divide it into different segments like historical, nature, cuisine, performance, so that users can chose and take a decision what they would like to do depending upon the available time and finance in hand. The last step in tourism is the return journey back home. This part is important because any personal touch, services, help provided to the users creates a long-lasting memory about the services provided. In many places as a custom, small mementos act as a stimulus for evoking those sweet memories of the tour and the place as such. The journey ends after the user enters home safely and with lots of sweet memories. The return journey needs to be planned equally well, because as users are bad in short term memory, they remember the last part of the tour more vividly than the first part. As it's all about the mental model which develops in users, the journey, destination, services, etc. should be in sync with one another, taking care that what has been planned and promised is strictly adhered to. Any extra elements provided to users should not fatigue them but should enhance their energy and excitement to know and explore more. The journey ends after users reach their home and put their luggage down in their own rooms, unpacks and keeps everything back to its original place, thus paving the way for another planned tour. It's important that this full circle is completed without any trouble as it comprises of people, services, places and products (Fig. 1.6).

1.11 Travel Modality

Users travel by different means. It could be individual mode (car) or mass transport (buses). In all these cases the user's body dimensions, should be factored in while designing the interior of such modalities. Apart from static body dimensions, dynamicity involving body movements along with luggage, etc. should be factored in as well so as to arrive at an optimized space for all different types of users. The human body dimensions need to be factored in and one needs to consider the database for anthropometric dimensions for working on this. The reach, movement pattern of users, ensuring comfortable sitting all needs to be factored in while designing different travel or transportation modalities. Apart from this specific information about the travel at regular intervals giving feedback to the users regarding the travel, its history, safety aspects and importance should be given in the user's cone of vision as they travel. This creates the scenario and sets the mood for travel for users.

Fig. 1.6 Travel thread from home and back to home. Photo by Tatiana Syrikova: https://www.pexels.com/photo/retro-globe-with-handmade-toys-against-gray-background-in-room-3932932/

1.12 Food, Photo Shoot, Explanation and Experience

Tourism as we have seen before is not just traveling to a place but is a larger picture. It consists of travel, safety, meeting people, food, shopping, relating what you see with the real world and ultimately feels enriched. To complete the collage the food should be of the user's choice, but explorations should be allowed so as to increases one's perception of the food. The food should look and smell good because these two senses work in tandem. Users are obsessed about photography while no tours as they want to lock every bit of their sweet memory forever to cherish. Food and the elements around are tangible elements which one can see, touch and feel, and the photoshoot is a medium to capture the moments. The experience one captures is to be cherished there itself and can only be relived later. Thus, exposure of user's that tangible and intangible element helps in eliciting those long-lasting experiences by forming a perfect framework for that unique experience generation. This is important in tourism and the reason why customers come back time and again to gain that experience once it's got empty. How do you generate that? Provide your users more than what they want, feedback at every step and the framework that stimulates all the senses eyes, ear, nose, tongue, touch, smell so as to create that unique experience comprising of all the senses. Add variety in between so as to break the monotony and create variation in presentation of tourist spots which holds your customers like glue to the tour and travel.

1.13 For Some It's the Destination for Some it's the Journey

Every customer is different when it comes to tourism. For some they enjoy the journey, others prefer reaching the destination fast, others it's only the cuisine and the culture which matters. Thus, from an ergonomic perspective you have to take care of the length of the journey, which should not be very long duration for elderly users. If the length of the journey is eight hours or more than a day's rest should be given for elderly users so as to recover from the fatigue. The journey in transportation should ensure that seats are comfortable with proper back rest and inclination so that users feel relaxed. If possible entire torso should be supported with support for the legs so that legs do not swell. In tours these counts toward customers comfort and satisfaction. Focus should be to ensure that there should not be back-to-back travel and adequate rest between. When one long journey ends give a night's break before embarking upon another journey. Thus, all touch point's needs to be factored with ergonomic issues starting from seat, visibility outside, information in the space, etc. You need to think of the challenged and the elderly as they are the ones needing special attention all through the journey. Specially access to places and vehicles, joint mobility, visual problems where the letters have to be big in size supplemented with auditory feedback. Floors to be textures so that chances of slippage are reduced.

1.14 Safety Aspects in Tourism

Tourism should be safe. If you tell users just to use some protective measures, gears, harness, etc. it might sound too harsh and boring and they might not adhere or obey that. The same way we refrain from using safety helmets, using seat belts in cars or as child play on the roads when our parents told us not to do so. So, when we don't abide by what we are told how can we expect that users would listen to our safety advises? So, the best way to ensure users follow safety norms is to "sell" safety instead of enforcing it on them. Explain to users that if they don't use the harness they might meet with an accident and injure their head and lead the life in bed. Once this picture is etched in the minds of the users, they are bound to obey it. The second aspects because users don't use safety products like helmet, gloves, etc. is that they do not fit the body of the user and are either too loose or too tight. This is at time accompanied by lack of ventilation. The cumulative effect is an extreme discomfort, which is why users do not use safety devices. So, they should be designed in tandem with user's body dimensions ensuring adequate comfort. The product should also gel with the existing human attire and not make the user look like an odd man out.

1.15 Snowball Effect in Tourism One Tour Ends and Planning for Another Begins

The good experience of one tour should lead to the genesis of the next one. Thus, as the tour comes to an end you can start pitching for the next tour to your users highlighting to them the specific features of the next tour. It's easy for users to imagine the tour characteristics and you mention to them the next tour with contrasting features it becomes easy to understand and register the experience. For example, if your users are just coming back from a tour of the mountains pitch to them the next tour to the sea shore, which is in stark contrast to what they have been exposed just now. But since they have experienced the mountains and sea side being opposite to mountains it's easy to "experience" the beauty of the place suggested and the tourists fall in love for it and book their slots for the next and subsequent trips and tours. Humans always perceive with reference to some existing experience (Fig. 1.7), both good and bad. It said that a burnt child dreads the fire. In tourism this is also true; the god experiences people always like to cherish.

1.16 The Problems with Excess Information: We Chew More Than We Can Digest

When you visit any fair, especially tourism fair you tend to visit all the stalls. Then you collect all the brochures from the stalls and bring them home, with the plan that you would be going through all of them. Ultimately what happens, majority of these brochures are untouched and they all land up in the waste paper basket. Thus, users have a habit of collecting more information than they can process. This is where you have to optimize tourist information else peopling get confused. A user in her/his entire life would like to cover every nook and corner of the earth. After a point they know that this is impossible and thus, they give up and look for specific information on tourism. This is where your help is needed. Present minimal information in short sentences and clear less detailed visuals which talks about the main features of a place. Restrict your features to 3 or at max 4 and not more than that. People are bad in short-term memory and would not be able to keep them in their head for long.

1.17 Key Points

A. Tourism and hospitality human factors demand a macro approach.
B. The macro and micro picture needs analysis
C. Keep customers, staffs and infrastructure in mind
D. It's all about providing the best services
E. You have to be different to win customers.
F. Remember your customer is always right.
G. Feedback with customers is the key to success.

1.17 Key Points

Fig. 1.7 One tour experience leading to another tour. Photo by Malte Luk: https://www.pexels.com/photo/man-in-brown-top-beside-railroad-2432299/

- H. The tourism canvas is important comprising of people, places, travel and accommodation.
- I. People look for specific information and have their own needs and wants while traveling which needs to be addressed.
- J. Everyone is different; hence factor that difference in tourism.
- K. Tourism is all about humans and starts from planning, execution and back home safely.

L. One tourism experience leads to another tourism based on user's good experience.
M. Keep an eye on user's comfort, safety, fatigue while traveling.
N. Information about tourism should be optimal else people get confused.

1.18 Practice Session

A. An elderly couple both of whom are wheel chair bound are planning for a trip to a pilgrimage site. It's a day trip and they are taking the help of a tourist agency. From the view point of the tourist agency, you have to identify the human factors issue that needs to be considered.

Directions:

1. First visualize the bigger picture and identify all the component.
2. Draw the pyramid and place the tourists and other components.
3. List down the tangible and intangible human factors issues.
4. Cross check to see whether the entire ecosystem of travel is now robust or not.

B. A tour and accommodation are to be planed for a group of school children. As a responsible tour operator, you have to identify the areas in this tour that demands human factors interventions.

Directions:

1. Safety foremost as they are children.
2. Variety as children get bored by one thing.
3. Different tourist spots as the group have lot of energy.
4. Engagement for them through games while in the hotel else they will disturb others.
5. Try to communicate to the through stories of the tourist spots rather than narration. It is more engaging.

C. Mr. Smith wants to visit the backwaters of Kerala in India. He is in Austria and wants customized tour planning for himself. He is 75 years old and suffering from Parkinson's disease. You have to use human factors principles in planning for his trip.

Directions:

1. A guide has to be arranged.
2. All the places he travels should have grab rails and textured flooring.
3. Guide should speak very slowly and communicate with gestures and short sentences.

4. Do not plan too many tourist spots in one day.
5. Make sure he gets rest after lunch and if possible, skip the post lunch session and let him rest.

D. Tim, Jack, May and Harry are four friends in their early 30 s. They want to go for a trek to the mountains for themselves. They are looking for a customized tour planner and operator who would plan and execute their adventure. You have to use human factors principles in help them plan and execute their adventure tourism.

Directions:

1. Adventure means users look for that excitement.
2. Safety aspects should be explained to all repeatedly.
3. Make sure the gadgets are in accordance with the body dimensions (anthropometry) of all users. This is very important.
4. Keep them informed about the possible dangers repeatedly so that they develop a mental model of the same and are prepared.
5. Inform them the availability of services, medical facility, etc.

E. Mrs. Sweat is suffering from a rare cardiac disease. Doctors in town are unable to detect or treat it. They recommend her to look for advanced treatment in some other cities but are not sure where to go. Mr. Sweat approaches a tour operator specializing in Medical Tourism. You have to chalk out their plan, execution and bring them home, through the application of human factors principles. Mr. Sweat her husband is going to accompany her.

Directions:

1. The couple needs to be presented with options along with cost so that decision making is easy.
2. Show them the accommodation available in the vicinity of the hospital in case they need to stay long.
3. Project to them the tentative budget.
4. Give them a complete breakdown of every cost.
5. Inform them what help can your tourist company render in case of any emergency.
6. Present to them the treatment protocol. They can mentally prepare themselves.

Bibliography

de Gois Leite CM, de Carvalho RJ (2012) The ambivalence of the work of the hotel managers: an approach to human factors. Work 41(Supplement 1):5668–5670

Latip SNNA, Latip MSA, Tamrin M, Nawi MZM (2022) The perspective of work human factors on employee task performance in hotel and tourism industry, Malaysia. Proceedings 82(1):7. MDPI

First Love with Accommodation 2

Overview

This chapter discusses the ergonomic issues in tourist accommodation. It takes the readers through different touch points for the guests when they arrive at their accommodation and the ergonomic issues that regulate their likings and disliking. The chapter also discusses how ergonomic principles could be used to make any room look impressive for the guests so that they like it at the very first go. The issues of crowd and how to deal and interact with the crowd during peak tourist seasons are discussed.

This chapter gives an overview of how to design different elements in the hotels and accommodation for the tourists and the staffs keeping in mind human body dimensions. The different movement attributes and how they play a key role in designing of the space and the elements in space have been discussed.

2.1 Accommodation: The Humane Touch: Lobby, Counters, Clocks

The reception is the first interface in the hotel with the guests and plays a very important role in generating the first impression of the guests. Human factors of the reception suggest that staffs are well dressed with clean attire and properly groomed so that they look "organized". The next comes is how they speak with the guests. A smile is very important along with an element of firmness else people take you for granted. The smile suggests (Fig. 2.1) that the guest can ask questions and the firmness is an indication that anything non-sense by the guests (guests trying to flirt with staffs) is a strict no. This very approach helps in drawing the line in between the guest and the staffs at the reception. The next direction that should be taken by staffs is greeting the guest in the traditional way of the place. Staffs should be selective about the choice of words for their guests. Always tell

Fig. 2.1 Reception staff at hotel with a smile is a welcome gesture. Photo by Cedric Fauntleroy: https://www.pexels.com/photo/a-woman-talking-to-a-receptionist-4266932/

people what to do rather than what not to do. So, the sequence of words should be positive first and then, if necessary, the negative (what not to do part).

Staffs have to do multitasking and this too amidst a very noisy environment. Thus, this can lead to early onset of fatigue, and thus micro breaks are warranted for staffs to increase their efficiency. To give a calm and soothing ambience to the reception plants for different types along with good illumination plays a very important role. If finances permit then artificial water bodies and fountains could be created ate reception which gives the place a soothing appearance and helps to ease the work done here.

When tourists first enter the accommodation after a long and tiring journey they have some expectations. First the place has to be comfortable like home, should be clean and friendly people with personal touch. It's thus important that the lobby is designed with ergonomic issues keeping these in mind. The entrance should be spacious which would give the users an impression of free and open space supplemented by few furniture pieces for waiting if needed. A few plants or an aquarium brings lots of peace to the mind and specially the aquarium acts as a source of attraction for younger tourists while the elderly are busy checking in. Remember your tourists are fatigued after a long journey and would like to retire in their respective rooms. The lobby is the first interaction with the tourists and should give that mental model of the place with a little insight into the major must see for tourists. The wall clocks with different timing of various countries are a value addition as it acts as a feedback for foreign tourists as to the time in their home country. This helps them to take a decision when can they call back

2.1 Accommodation: The Humane Touch: Lobby, Counters, Clocks

home so as to inform their near and dead ones about their safe arrival in their accommodation.

The counter or the reception counter should facilitate direct eye to eye contact with tourists and should not look like a hierarchical arrangement between the tourists and the hotel staffs (Fig. 2.2). Rather it should be projected as a defined zone integrating in the larger space. The material and finish of the reception plays a very important role to convey this feeling. Normally uses of wood for counters are better perceived as image of elegance and aristocracy along with the logo of the company. The people at the reception should be well groomed with smiling face, because the first human touch is very important in creating a lasting impression among the tourists and make them feel at home. This first mental model creation is important and if successful the tourists would return to your hotel time and again in future. The subsequent interaction might not be that pleasant and it would still work. The lobby should give a snap shot of the accommodation and a glimpse of the facilities it contains. This helps travelers in decision making in case some changes in the travel and accommodation planning are required.

Fig. 2.2 Direct eye-to-eye contact at the reception plays a very important role. Photo by Mikhail Nilov: https://www.pexels.com/photo/a-female-tourist-talking-to-the-female-receptionist-of-the-hotel-7820311/

The reception counter should facilitate eye to eye contact of the staffs with the guests. The counter top should be at the guest's elbow level so as to facilitate easy standing at the counter with the hands over the counter surface. If necessary, the guest can also fill up form, sign papers, make payments, etc. The counter in front of the guest should have a concavity with place for placing the foot inside, which would help guests to stand in style while folding their legs at the knee (Fig. 2.3). If the front part is flushed then standing becomes difficult and guests stand away from the counter and the counter top as there are no place for the folded knee and placement of the feet. Guests at the counter have very little attention span and hence pen, paper, forms everything should be kept within the guests reach so that they don't have to ask for it. Some drinking water provision in the form of small packaged drinking water should be kept as a gesture and creates a good image of the hotel to the guests. All the activities involving document checking, photocopying of guests document should be done in the line of sight of the guests so as to give them feedback as to what is happening to their documents and the same should be handed over to them as soon as the photocopy and document checking are completed. The room key in whatever form it needs to be handed over to the guest with proper label of the room number.

2.2 Locating the Place in the Actual Context

For solo travelers locating the accommodation becomes difficult because of different reasons (Fig. 2.4). The traveler is already under stress and is in an uncertain situation as to when would he/she reaches the accommodation? Would she/he be on time, what if he/she is unable to reach and so on? It's here that information related to the accommodation needs to be carefully put up to reduce this uncertainty among tourists. There is no point in putting up an information only on the premises, because users would like to know about it much before. The information on the premises is important but only acts as labels that this is the accommodation where the tourist is supposed to go. But as the tourist/tourists move they need to be given some feedback about the accommodation as they travel so that they know that they are in the right direction. If on foot then the information like "you are 500 m from XYZ hotel" proves beneficial and acts as reconfirmation and reduces uncertainty. While giving address of the accommodation its better if "landmarks" are referred to as information retrieval with reference to landmarks becomes very easy. For example if you tell your client walk for 100 m and you see a burger shop, from there turn left and walk for some 10 m and you come across a church. The accommodation is next to the church. So in essence the buildings that you are talking about are "landmarks" with reference to which it's easy for users to locate the accommodation with ease. If the same is in a crowded place, merged with other building then you have to make it stand apart from others. The building can be made to stand apart in different ways either through color, texture or through specific architecture which stands apart. If the accommodation is painted in a bright yellow for example, it might stand apart (provided other buildings are

2.3 The First Interaction Wow

Fig. 2.3 Guest standing at the reception counter. Photo by Mikhail Nilov: https://www.pexels.com/photo/a-guest-talking-to-a-woman-at-the-reception-area-7820376/ Figure

not so). If it's shaped like a ship then also it stands apart like an odd man out, saying as if "I am here".

2.3 The First Interaction Wow

The first interaction with any place or people is very important and leaves a permanent mark in people's life. Humans are very bad in short-term memory and are very selective when it comes to remembering. One of the human characteristics is that they give lot of importance on the first interaction. This interaction could be with people, place, situation or a particular experience in total. Thus, here if you are able to win the hearts of your user, you and your company can create a lasting impressing among the customers. After this even small errors in services in the hotel are often overlooked by the customers. The first interaction can be facilitated mainly by good human behavior coupled with personal touch. Say a nice "hello Sir" but if this is coupled with "did you have a good night sleep last night sir?" This

Fig. 2.4 Tourists trying to locate accommodation through signage system on the road. Photo by Davis Sánchez: https://www.pexels.com/photo/woman-standing-beside-alto-signage-2006953/

extra statement adds flavor to the dialogue and elicits that experience among the customers. Tangible elements on display like a cozy recliner chair, where tourists can lie down and relax also can elicit that comfort and calmness which elicits that wow factor in the tourists (Fig. 2.5).

Personal grooming of the staffs is an identity of the hotel and adds value to the hotel ambience. Staffs need to wear clean dress as this conveys an overall cleanliness message to the guests about the hotel. The color of the staff's attire should reflect the company aim and objective and should also help the guests differentiate the allocation of function among the staffs. Staffs should groom properly taking care there are no excess makeup which makes them pop up from the ambience and stand apart in the oddest manner. In case staffs are suffering from infection which manifests them in the form of sneeze, cough, etc., then such staffs should be put on sick leave and if necessary, they should be designated to some back-end tasks but not at the reception. Guests always give too much emphasis on what they see first and if they see staffs at the reception with infections, they immediately draw a conclusion about the lacuna in the health and hygiene of the hotel. Staffs

2.4 Staffs Look and Feel

Fig. 2.5 First interaction and the customer's wow! Photo by RODNAE Productions: https://www.pexels.com/photo/a-female-front-desk-receptionist-smiling-to-a-client-7755676/

at reception should not eat, or drink in the vicinity of guests as that creates a very bad mental model of the services of the hotel to the guests. This mental model (mental image) suggests a lack of hygiene in the hotel as staffs eat and touch different documents with the same hand leading to transfer of food particles through the document to the guests.

Luggage handling is loading lifting. The staffs of the hotel associated with luggage handling should take care of their back. Load should be kept close to their body while lifting. And while lifting the knees should be bent keeping the back straight so as not to injure the back. Repetitive movements should be avoided at all cost. Heavy luggage if they do not have wheels should always be handled on trolleys to prevent injury to the back (Fig. 2.6).

2.4 Staffs Look and Feel

How would you feel if the person in any hotel reception is standing in a Bermuda, bare body and talking to you while brushing his teeth, with the tooth paste oozing out through the corners of the lips? Very irritating isn't it? Compare this with the person at the reception well groomed, clean shaved and talking to you in a very polite manner (Fig. 2.7). You feel much better and it creates a better image of the accommodation or the services which is being rendered. Humans are social animals and always expect that personal touch which is achievable through an

Fig. 2.6 Luggage trolleys for carrying heavy luggage. Photo by Mikhail Nilov: https://www.pexels.com/photo/a-woman-talking-picture-at-the-reception-area-7820313/

integration of better personal grooming coupled with excellent communication skills with customers. Thus personal grooming of staffs not only generates a better profile of staffs but also helps the staffs integrate with the ambience of the accommodation and in totality presents an ambience comprising of space, element, lighting which are inanimate along with well-groomed people who are animate who add value to the space in terms of breathing life in the space. This personal grooming thus has to be in tandem with the theme of the place and the services being offered else it would be an odd man out.

For the reception the grooming is, for example, full sleeve shirt and dark trouser with tie. For the chefs it's the white chef's coat with the cap. For service the grooming is again full sleeve shirt and dark trouser at times coupled with gloves and apron when it comes to cleaning of toilets. Thus personal grooming gives specific identity to people in the space and also communicates to guests the division of services among different staffs. This adds to the overall ambience of the space where there is seamless integration of people, services and elements of space. As users scan from top to bottom it's very important that the upper part of the body should be groomed very well. This includes the hair, face, shirt, tie and then the rest. As people scan you their least focus are on your shoes. That does not mean that you are not going to wear shoes!!!

The experience of someone taking care of your luggage the moment you arrive at the hotel is unique. Bell boys play very important role in doing this and help in screening the same. This task again should be done in the presence of the guests

Fig. 2.7 Well-groomed staff generates a better image of the place. Photo by Ketut Subiyanto: https://www.pexels.com/photo/woman-in-gray-t-shirt-checking-in-and-smiling-4907441/. Extra Counter with concavity: Photo by Pavel Danilyuk: https://www.pexels.com/photo/men-buying-movie-tickets-7234477/

as they are jittery about the fate of their luggage and would never like to lose sight of the same.

Inside the hotel once the check in is done guests are lost in space where to locate their rooms. Bell boys act as guide and carry the baggage to the room. These gestures of guidance to the room help the guest. Moreover, the bell boy opens the room and it's presented to the guest in a manner that he likes it at first sight. The lights are switched on, facilities; shown, complimentary elements are indicated. If it's daytime, then the curtains are partially opened to give the guest a view of the world outside. As many guests are claustrophobic in elevators, they should be accompanied in the elevators so that they do not panic in the closed space. This gives the guest a feeling of being at home and being taken care of and prepares them for a comfortable stay in the hotel.

2.5 Importance of Check-In and Check-Out Counter

The first interface with the hotel is the staffs at the reception, and they leave a lasting impression in the peoples mind. Tourists always give priority to check in and check out in a hotel. At the time of check in the experience generated is important and has a lasting impression in customers mind about the place. You need to be fast and at the same time efficient in checking in your guests because they are tired and have come to their second home where they would like to stay and enjoy.

When guests check out they have already had a good or bad experience of the place. If it's good you have to retain it, if it was not good you have the last option to rectify it and give them a better experience. This is where it's better to segregate the check-in and check-out zones at the reception counter. This helps in dispersing the tourists with two different objectives. One group, tired, uncertain looking to check in. The other group, relaxed, looking at check out. Both the groups are important for you, but they differ in their mental model.

The group which is checking out needs to be give service fast, and at the same time ensuring some last minute services in the form of asking for their feedback or even giving a small memento as a remembrance of the hotel and even advertisement for the same. You can also provide complementary drop service because just like the first interaction the last interaction is very important as your guests have a very bad short term memory they remember it as a part of their journey.

There has to be a seating area, for those guests who are checking in and out and this area should be common. The reason is that it facilitates the interaction of two types of guests one checking in with those checking out and in most cases the experiences of those checking out in the hotel out are shared with those checking in it answers many of their queries. Thus those checking in waste less for their times as majority of their queries are answered.

2.6 How to Ensure Proper Crowd Formation

Guests especially when in large numbers crowd at the reception desk and area, and this becomes chaotic as such. You cannot force them to be organized but just can request them to come one by one. But what if this mechanism doesn't work? How do you control the crowd? This is where ergonomic and design could come to your rescue. People in general prefer eye to eye contact and coming in close proximity to the person they wish to talk to. If you can facilitate this through design, they would opt for it.

One way is to give the counter at the reception a cavity and a bulge. The receptionist stands inside the cavity at the other end and the guests would for obvious reason try to fit into the concavity so as to be in close proximity to the receptionist and in direct eye to eye contact. Now if there is a crowd, people would normally try to stand closer to the receptionist inside the cavity in the table (Fig. 2.8), so as to come in close proximity to the staff as that facilitates belter interaction. No one would stand at the convexity because in that case the distance between the tourist and the staff increases and this does not facilitate better interaction. The guest standard away can also see the other guest in close proximity to the staff. So automatically that guest tens to stand behind the one standing in the concavity and looking at this others also follow the same rule (herd mentality of crowd! Tend to do what others do!) and automatically a nice queue is formed without any enforcement. If you want the crowed to be further organized, then belt post (Fig. 2.9) can be arranged which indicates to the guests where and how to stand and deviation from that is not possible because of the physical barrier.

If after taking so much of care you find that overcrowding is happening then stagger the seating zones all over the place. Tourists in general are very active and tired and prefer seating rather than standing. In that case majority would prefer taking a seat (especially when traveling in a group) and only a few representatives would stand at the counter. These seating arrangements should be away from counter and thus automatically the crowd density at the counter or relevant places in the path of movement of people and material could be automatically reduced.

Fig. 2.8 Crowd control through design of reception counter. Photo by Pavel Danilyuk: https://www.pexels.com/photo/men-buying-movie-tickets-7234477/

Fig. 2.9 Crowd control through design at a counter. Photo by Lachlan Ross: https://www.pexels.com/photo/people-gathering-near-counter-at-festival-6510339/

2.7 Attention Span of Guests and How to Deal with It

The attention span of your guests is very short and at times they get angry for trivial reason. There are many different ways to deal with it. The problem happens when guests look for a service and fail to get it or any feedback related to it as to when it would be addressed. For example, your guest gets up in the morning and has the habit of taking black coffee along with cookies. This practice they have inculcated at home and expect the same in your hotel. So before your guests arrive its better if you can get a snap shot of their likings and disliking through a small questionnaire which helps you in gearing up before they arrive. In that case such services and arrangements could be made in advance and provided once they step in your hotel. The second approach is right information in the right place. If the room contains the information (very brief only 2 to 3 lines for each services) related to different available services that gives a feedback to the guests and help them to take a decision of which one to opt for and which one to not. If you are able to address maximum queries of your guests they won't look for that information. If you inform them that there is a tea and coffee maker in the room with a box of cookies then they can make it and have it on their own without having to wait for it to be served.

Even after making all efforts you might find some guests who would crib for service and shout when they are not provided. People crib when they have lot of free time, and this is true when people are on holidays. So if you can keep them occupied in some activities like water sports, games, or anything which is interactive, that is requires repeated input and output then it keeps their short term memory loaded and busy and thus they do not have time to think of what they need and thus crib for it. You would notice that once you engage them, they would stop cribbing and everything would be in place.

2.8 The First Impression of the Room

Users exhibit certain characteristics when they check in to a hotel room. The first thing they check is the toilet to see how clean it is because most of the tie the first thing the guests do is to relive them in the toilet. The next their focus goes on the bathroom and the bathing area. The biggest challenge is the shower and whether it can dispense hot and cold water in case you need to have a shower. Guests have no clue how to get this optimum mix of hot and cold water, and most of the time the experience is pathetic and they get jittery. Next their focus goes on the space available in the cupboards whether adequate or not. After this the bed, how big and comfortable it is. The sofa set and other furniture elements catch their attention next and the eyes scan around for anything to eat and drink including water. If there is one then the question arises complimentary or paid. You have to be clear and provide precise information on what is complementary and what is not. Next the view of the outside world, your guests would love to have this and possible one of the most memorable aspects of your guest as they stay in your hotel.

In essence when guests enter the room the eyes scan the space in totality with the space and its elements (Fig. 2.10). The liking and disliking is generated here itself. Thus, make sure the room is clean, all elements in space in place and properly oriented, lights are on and first remove the curtain as you step in the room so that guests get a view of the outside world. This is important as it removes the feeling of claustrophobia if people can communicate with the outside world.

2.9 Elements that Makes the Room Beautiful

The human eyes scan from top to bottom and then from left to right or right to left which is cultural. Thus, the eyes travel to the nook and corners which should be clean. The places behind photo frames gathers dust and needs cleaning. Guests are curious about the toilet and its cleanliness, and thus it should be maintained clean with the lights switched on so that the toilet and the bathroom is visible. The bathroom should be absolutely dry as the guest would first enter with shoes, and no way should there be any incidence of slips, trips or falls which should be enough to make him jittery. The next element is the mirror where the guest sees himself and hence this should be clean as well (Fig. 2.11).

For your guests the room should look beautiful else they do not feel that attachment. This is where that "home touch" is needed. A few paintings on the wall, nice sofa set, a desk /table for doing some writing task, a clean and cozy bed. This is

34   2 First Love with Accommodation

Fig. 2.10 Room as the guests visualize. Photo by Pixabay: https://www.pexels.com/photo/bedroom-door-entrance-guest-room-271639/

Fig. 2.11 Guests looking at the bathroom mirror. Photo by Andrea Piacquadio: https://www.pexels.com/photo/woman-in-white-tank-top-while-looking-herself-at-a-mirror-3811809/

what they get at home. Over and above some facilities like tea maker, a small fridge, some complementary snacks along with some reading materials like the newspaper and magazines makes the room beautiful. Cleaning and maintenance of the room plays a very important role as well if the guest is staying for long. Fresh flowers in the room specially the seasonal ones also play important role in making the place beautiful. A bowl of fresh seasonal fruits alongside augments the beauty of the room further by bringing in the fresh lease of natural element inside the room. Guests always love to have a slice of nature within their rooms. Lastly room fresheners can leave a lasting impression in your guests mind, provided it's of a specific variety and not available elsewhere. It creates a brand and an image of the hotel and the stay of the guest which they cherish for long. This image thus created (mental model) can be used to retrieve a; set memories of the guests hotel stay.

2.10 Humans and Room Ventilation and Illumination

Rooms should have facility for heating, as well as cooling. Provision for fresh air from outside also adds flavor to the room as guests prefer a slice of nature in their room. Ensure that there is adequate illumination in the room and specially provision for low illumination during night so that guests can navigate to the bathroom. Illumination has a relationship with people's mood, and most people are not comfortable with low illumination level inside the room if they are staying the whole day. The entry of natural light is also equally important which also elevates people's mood.

2.11 The Bed

The bed in the room should be low enough for guests to easily access and get down with ease. The length and width of the bed has to factor in not only the height and width of the guests but some extra allowance to it. The reason for this is that when you sleep your hand moves above your head and you move your legs and body. Thus the body moves a little up and down. So allowance needs to be added to this while calculating the bed length. Similarly when you turn to your sides while sleeping you roll from your initial position and thus the bed width needs to be calculated not by considering the body width but some extra allowances as well. Some people use pillows between two legs. So the width of the bed has to accommodate for that as well else you guest or the pillows or both might fall from the bed!

While making the bed for your guests house cleaning staffs need access to the bed, and thus the height of the surface of the bed should be at least a little below their elbow height (Fig. 2.12). Additionally there should be space for the staffs to move all around the bed not only for making the bed but also for cleaning the room and the space around the bed. For calculating this space, you may take the

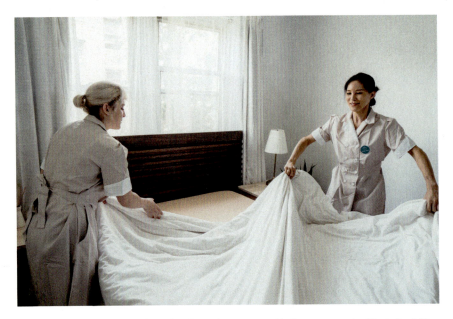

Fig. 2.12 Bed making and dimensional requirement considering movements. Photo by Liliana Drew: https://www.pexels.com/photo/housekeepers-changing-sheets-in-hotel-room-9462740/

reference of the hip breadth of the service people. It's better to take the fat person as the reference and in that case the thin person would have no problem. Keep in mind that the staff making the bed has to bend and move, so make sure you keep extra allowance for that as well.

2.12 The Table and Chairs

The table height should be fixed at the elbow height of the user so that writing becomes easy. The area of the table surface is decided by what type of task you want, what elements you want to fit. If it's just your laptop or a note book then the area traversed by the arms extended from the body could be taken an s a reference for deciding the surface area. The chair height should consider the height of the popliteal zone from the ground for the lower values, and the width the hip breadth and the higher values. The depth of the chair should factor in the length of the zone behind the keens from the buttock.

Above this factor in the fact that people would have to move the chair backwards for getting access to the table so place has to be kept for that.

2.13 The Bathroom and Its Elements

The bathroom dimensions should factor the height of the sink with reference to the elbow height of the shortest person while standing. Similarly the latches, towel rack soap rack, etc. should all consider the shorter guest as they are all issues of access. The mirror should be placed such that it should be able to accommodate users of different heights. The bathroom is normally for single person use. As it is for single person, all calculations should be done with reference one person only.

2.14 Bathtub

The bathtub is again a place where guests relax in lukewarm water. They are normally in a semi-sitting posture. The bathtub should be able to accommodate this posture of the user. It's not necessary to consider the height of the person in this case as that would lead to unnecessary material wastage. The width of the tub should consider the hip width of the user with some more allowances to it. The other element which needs to be factored is the height of the bathtub because it has to hold water. Here one has to be careful because the user has to climb on to the tub and also climb down while it's wet. Thus the edge of the bath tub should be at or a little below the knee joint of the user for facilitating this movement. Your consideration here should be the lower percentile or shorter users.

2.15 Remote for TV and AC

In any hotel the user is confused with TV and AC remote. This is because they are used to a different layout of both at home and they do not have enough time to learn the new layout, and hence it becomes frustrating for them. The ways out from this problem are either to have a remote with very few options of just channel change and volume change. In case of AC remote with just the option of increasing and decreasing the temperature and stopping the AC. Else complete pictorial depiction of how each is to be operated with minimal text usage should be kept in each rooms.

2.16 Window Curtain

Window curtain in the room are an important tools for your guest. It's for their privacy, and it's to be opened for getting a view of the outside world. Thus control of window curtain should be within easy reach of guests. It's better if guests can open and close curtains while in bed because only after lying down people realize the exact extent to which the curtains needs to be opened or closed. Else if it's at a distance from the bed guests have to get up often either to shut off external light or to ensure that they can see the outside world in the morning while in a half awake state in their bed which is a unique experience in a new place.

2.17 Light Control from the Bed

In the hotel room guest would need the lights at any time. They are not familiar with light switches like at home. Thus light switches at least a few should be absolutely near the bed and in the expected quadrant of the persons reach. Normally near the head of the user so that while stretching the arms it could be activated to switch on and off the lights. The switches should have fluorescent citing so that they are visible in pitch dark room as well.

2.18 The Bathroom Direction at Night

At home you can easily navigate to the bathroom without swathing on the lights. You develop this art through repeated practice. In a new place like hotel this is not possible. But people in deep sleep when they get up at night for going to bathroom often forget this as tends to trip and fall. This is where floor lighting in rooms are must which would help guests at least see the floor and thus navigate to the bathroom in case they don't want to switch on the room light.

2.19 Replenishing Tea, Coffee, Toiletry, Towels and Soaps

All complementary elements need to be replenished even before they are asked for. This sends a good message and creates a good image. Any complementary product which lies blank is perceived as negligence and an indication of poverty on part of the hotel by the user. Users take or make judgment very fast with whatever is available to them at hand. Spare bed sheets, pillows and blankets should always be provided as guests need them very often as their needs vary.

2.20 Means of Detecting Missing Elements in Room

Any extra organization or disorganization in the room is an indication that certain things have gone missing. Look out for small elements as it's difficult for guest to steal large elements. Any blank or empty space in the room after a quick scan also indicates missing elements in the room. If rooms in the same price range have the same organization of elements in the space, then staffs develop a typical pattern. Thus, any change in this pattern in any of the rooms could be easily detected indicated some elements missing.

2.21 Key Points

A. Human touch in hotels is the most important creating a lasting impression in guests.
B. Tourists need to be engaged in hotels with activities else they start cribbing for trivial matters.
C. The placement of the hotel can be crucial for locating it in space.
D. Specific information in brief to be provided to guests in writing in the rooms for the guests.
E. Crowd control should be by design intervention and not by enforcement.
F. To ensure home touch use specific elements inside room.
G. Make sure guests get an outside view from the room for preventing claustrophobia.
H. Human body dimensions need to be considered in room design.
I. For access consider the lower value and for clearance consider the higher value for body dimensions.
J. All elements in the room should consider people movement both the guest and the staffs.
K. Remotes should be universal and have minimal options making it easy to understand.
L. Curtains and switches should be within easy reach of guests.

2.22 Practice Session

A. A hotel is located on a hill top. Guests traveling for the first time are finding it difficult to locate it in space. What human factors directions can you suggest for helping your guests?

Directions:

1. Think of proper directional signages on the road.
2. The signage is an information, and it should be strategically located specially at bends, bifurcations.
3. You might need to repeat the signage at intervals.
4. For people on vehicle the text should be minimal.
5. The color of signage should be of contrasting color with the surroundings so that it stands apart.
6. Place the signage calculating the eye height and visual cone of the users while walking and while in vehicles.

B. You have a group of young guests from a company manufacturing mobile phones. The guests are always complaining of trivial things. What human factors approach would you opt for them?

Directions:

1. If people have no work, get bored they start complaining.
2. Think of engaging your guests.
3. Select activities they like, if possible, group activities.
4. Advise them for local shopping for some exotic products of the place.
5. While traveling people are on shopping spree, encourage them.
6. Give guests information on what the hotel provides, this feedback is important.

C. Your guest is an artist from another country. What human factors features would you incorporate in your room so as to impress your guest in the hotel?

Directions:

1. You need to match the mental model of your guest. The way they think and expect.
2. Some good art work are a must in the room.
3. If possible, promote work of local artists.
4. An artist would like to see how a particular art is produced, depict the steps through visuals.
5. Keep information about art exhibition, art sale, happening during the period.
6. Keep information and contact of local artists.
7. Information on some local art school and colleges should be kept in the room.

D. There is a huge crowd at the hotel reception. Guests are checking in and out. How would you control this crowd without letting the guests feel that it's being enforced upon them.

Directions:

1. Focus more on those checking out as they are in a hurry.
2. For those checking in, send them for snacks or beverages which is complementary.
3. Arrangement for an audio-visual display of the must-see places is also an option to engage those checking in and waiting.
4. Some information on local products and local market for the new guest would also engage them.

E. You have to define the dimension of compact room in a hotel for budget travelers. Define the ergonomic directions and approach for the same.

Directions:

1. The elbow-to-elbow distance in the reference point.
2. With the forearms folded move a step back, sideways and front, this would give an idea about the clearance required.

3. Lie down on the ground and check for clearance between elbows and below the legs and the head.
4. Start calculating the room dimensions from the database of body dimensions.
F. A big single bed room needs to be converted to a double bed room. It needs to have a bed, table and a sofa. Discuss the ergonomic directions for this arrangement.

Directions:

1. Calculate the free space available after subtracting the space taken by the furniture.
2. Calculate the work space envelope for two people by taking the elbow-to-elbow distance and the fingertip distance after spreading the arms at the sides.
3. Add extra allowance for movement. For this check for the type of movement in the room.

Bibliography

Ismail FH, Osman S, Rahman FBA (2021) Human factors kitchen: a better place to work. Int J Acad Res Bus Soc Sci 11(13):43–53

Mahadewi NM, Irwanti NK (2016) Image Bali as an ergonomic island destination. In: Asia Tourism Forum 2016-the 12th Biennial Conference of Hospitality and Tourism Industry in Asia. Atlantis Press, pp 156–159

Host and Hospitality

3

Overview

This chapter gives an overview different dimensional aspects in hotel and dining. How human body dimensions are used for design of space and products are explained in this chapter. Readers are given an overview of the ergonomic issues associated with dining. Different aspects of food, especially food design and presentation are discussed here. The crowd flow during dining and special occasions and the different issues that guests face and complain and their ergonomic angles are discussed. The chapter touches upon the application of simple ergonomic principles in generating that desired ambience for the guests.

3.1 Fitting Your Guests and Staffs

Mr. Neelu was traveling to a country "Desire". When he arrived there, he checked in to a budget hotel. He was shown his room, but as he was about to enter the room his head bumped against the door frame and he was in pain. The staffs of the hotel came forward and made him to lie down on the bed. Neelu could not lie down comfortably as the bed was too small, and his legs were out of the bed and in fact it was too narrow for him. However, after applying ice pack Neelu recovered and he thanked the hotel staffs. He went to the bathroom and thought of having a shower. The shower was fixed to the wall and he found that his head was almost touching the shower head. Wherever in the hotel he felt as if he was a misfit as all the furniture, space all appeared to be for guests who are shorter than him. Neelu came from a country "Lombu" where people are on an average taller than the rest of the world.

   Ms. Sundori has just joined a hotel in Lombu in the room services. She is from a country "Happy". She is enjoying the place and her work. But she has a peculiar problem. She finds that everything is too big in the hotel. The doors are very big,

© The Author(s), under exclusive license to Springer Nature Singapore Pte Ltd. 2024
P. Mukhopadhyay, *Human Factors in Tourism: A Layperson's Approach*, Design Science and Innovation, https://doi.org/10.1007/978-981-99-7066-7_3

the table top is very high, the door knob is difficult to reach. She even finds that while cleaning the rooms she is unable to reach the mirror in the bathroom and thus has to climb on the wash basin to clean the mirror. For cleaning the ceiling of the room with the normal broom with long handle, she is unable to reach and has to attach another handle to do the task. She comes from a country where people in general are of short height compared to the people of Lombu.

A group of tourists from Lombu (where people are very tall) is visiting the country Happy where people are relatively short. They plan to take a boat ride in one of the famous tourist spots where the boat takes them through marble rocks. When they got into the boat they found that the seat was unable to support their buttocks and even two people could not sit comfortably side by side. They were so uncomfortable as they could not sit in the boat properly and thus could not enjoy the beautiful boat ride.

The above three incidences are typical examples of mismatch in body dimensions of users (guests and staffs) when it comes to any place of accommodation. This is a big challenge. You never know who is going to come to your hotel, either as a guest or as a staff. The hotel and its different elements (furniture's etc.) has to be comfortable and moreover accessible to everyone irrespective whether they are tall, short, fat or thin. If this is not factored in then people would not be comfortable in your hotel and in this era of cut throat competition in the tourism and hospitality industry you will lose customer. This is where the study of human body dimensions (called anthropometry) becomes handy. So, while designing spaces, furniture's, beds, etc., one has to consider the human body dimensions.

The calculation of space and different element in space has to be done in steps else. Experts have already created for us some data base for different human body dimensions (anthropometric data base). We have to learn to use this data base for designing. There are data base for different parts of the body for the male and female population of different parts of the world.

If you look at any anthropometric database you would notice that the data for each body part is represented in the form of percentile. It means a representation out of 100. So 5th percentile value means the smallest value and 95th percentile value means the largest value. While you pick up these values you have to remember that when it's a question of reach, for example the height at which the door latch should be fixed, we normally become a little biased toward the lowest or 5th percentile value and then optimize it for all This is because if the shorter person is able to access it the taller person would have no problem but the opposite is not true. Similarly, if we are deciding the height of the door then we normally become a little biased toward the taller or 95th percentile value and then optimize for all. This is because if the taller person is able to cross or enter the door without hurting his/her head the shorter person would have no problem. We need to add movement of people, clothing, shoes, hand gloves worn to these values as well. The reason is that the values have been taken while keeping the users in resting condition. But in space people move they are not static.

For doing this calculation follow some simple steps. First identify the dimensions of the space which you would like to calculate. Then identify what parts of the body will touch those part in space or the elements in space. Next go to the table in the data base which has that body dimension. Ask yourself whether this is a question or reach or access. If reach start with lower percentile value and if access start with higher percentile value. Remember human body is disproportionate and is a mixture of different percentile. Thus, my height could be 95th percentile but my shoulder breadth could be a little less at 75th percentile. Thus, always refer to individual body dimensions.

3.2 Injury of Staffs and Prevention

Staffs need to be physical fit for preventing injury. Job rotation, use of assistive devices, could be used to augment heavy load lifting and repetitive movements. The main reason why staffs suffer from injury is mismatch in the dimensions of space and tools they use. It could be that elderly person in poor physical state are working and they are prone to injury. Or it could be that people are exerting more force than their capacity and hence this is leading to injury. Working in awkward postures are can lead to injury. All staffs should have adequate rest to recover from micro wear and tear in the tissues of the body, else they would be prone to injury.

3.3 The Dining Room

The dining room in any establishment associated with hospitality industry plays a very important role as they are one of the major places in the establishment which showcases the industry/establishments' identity. The dining room is a space where people would move, dine, enjoy, talk and finally go away. Thus, the room should be designed in a way to accommodate the stipulated number of people. So how do you decide whether the dimension of the room is apt for the number of guests that you have invited in the space? Just perceive the space in two views, from the top (called plan view) and from the sides (called elevation). Your visualization from top would give you an idea about how many people could be accommodated considering the different furniture. Your visualization from the side would give your insight into the height of the table, chairs and other elements and whether people would be able to access the food or not. Normally if two people are sitting or standing while dining their elbows should not touch and there should be ample space for them to move without touching each other's elbow. Similarly, the dining table height should be around the elbow height of the guests so as to ensure that they dine with ease. When people eat ensure that on the table you keep adequate space between two person so that while folding the forearm and stretching the two upper arms on the sides, there are adequate space between two people (Fig. 3.1).

Your dining tables should have arrangement for people to sit face to face in groups for those who travel together but should also have provision for solo travelers who prefer sitting alone without having any strangers in front of them. Some seating's should be facing the windows with single seats for such solo travelers. If

Fig. 3.1 Adequate space between people while dining. Photo by fauxels: https://www.pexels.com/photo/sharing-cherry-tomatoes-3184188/

guests have to get their food from the table, then care should be taken to keep the seating areas away from the food zone. There should be consideration for people moving around the food table and you should factor this dynamicity to ensure adequate clearance between the food table and seating areas. Thus ideally the food serving table on one side toward the wall is a better option so that people can come, take their food and move to their seats for eating. Users prefer food to be taken their own way and hence let them have their choice for taking their own food instead for serving them on their plates. This plan can change depending upon the situation.

3.4 Elements of Food Design and Layout

Every food should look good, fresh and appetizing along with a good fragrance. Artificial room fresheners, or insecticide should be avoided at all cost as they mask the food fragrance and give rise to a peculiar smell which confuses the guests as to the food type. The layout of the food should be on clean vessels, and this is where metallic or glass cutlery should be used to convey that essence of hygiene. Food should be covered to emphasize to the guests that utmost care is taken to ensure its protected from insects and flies like is done at home. But here the cover should be of a transparent material so as to give feedback to the users as to the type of food the vessel contains. Your food should convey certain information to the users to reduce their uncertainty about what exactly they are going to eat. The appearance of the food should convey what it is. A confirmatory measure or reconfirmatory measure what users look for when it comes to food it's prudent to put the ingredients of the food as a label near the food. This measure helps users take a prudent decision and removes unnecessary time wastage in the queue along with other guests.

Users while eating look for certain critical information like vegetarian, non-vegetarian, vegan food. These most important classifications should be available to users and redundancy should be used to communicate the message loud and clear. There should be separate tables for each category along with labeling and possibly usage of the icons and color for vegetarian, non-vegetarian in tandem with population stereotype should be used. There should be the choice for separate dining areas for different categories of food, unless users themselves prefer to dine together. Always ensure that used plates, leftovers should be removed for users' sight as they create a bad mental model (image of the situation in the brain) which makes dining difficult for many.

It's known that food should be arranged in the sequence in which they are to be taken. For example, start with the starters and end with the desserts. But there are users who prefer to concentrate on only a specific category. There are again users who would like to take a second serving of a specific category. The layout of the food in such cases can open up the service table on both sides so that the user group could be divided into two different types without creating any bottleneck, where one group which would prefer to try all foods and the other targeting only a specific category or taking a second serving of the specific category. Those taking a specific food for the subsequent serving can use the other side of the buffet table for that purpose.

White fabric on tables gives an impression of hygiene, but once food spills its looks prominent and ugly and might create an odd man out like scenario leading to a distorted mental model among users who might not feel comfortable. These vessels holding food should be broad enough so as to account for users hand tremor and lateral movement. The serving bowls can be kept on small pieces of fabrics which. Could be instantly replaced in case food spills over them. This is natural as they are already holding a heavy plate in hand which leads to static load on the non-dominant hand and the task becomes tedious when it comes to serving oneself. Having the plates of a material of lighter variety might reduce the amount of static load on the forearm muscles.

3.5 Equipment Handling

The dimensions of the vessels holding the food on the table are decided by the volume of the food to be held. Vessels should never be filled up to the brim as during movement spillage could happen. This also makes lifting of food on one's plate difficult because the dominant hand has to handle the long and heavy spoon while standing and this act while standing might prove difficult for those who have problem with their knees and are unable to stand firmly on the ground. The plates and other cutlery's dimensions are fixed taking into account the amount of food it should hold. It's here that human hand dimensions like fingertip diameter, semi-precision grip by the fingers, palm breadth for bigger vessels on the table should be factored in to ensure proper fit with the human hand for the guests as well as for the staffs who handle them for the purpose of cleaning and maintenance. The spoon dimensions should map the oral fissure of the users so that they go inside the mouth with ease and does not get stuck at the entry point of the mouth. Spoon handles should ensure proper pinch grip and should be textured for proper tactile feedback and command over the spoon. As the spoon augments the capacity of our fingers!!

3.6 The Buffet Chaos

The buffet often faces chaos because of many reasons. If specific food gets over, people break the que in between to get a second serving, or guests only willing to get a specific food type can create chaos. To control this multiple buffet tables at four corners of the room can be arranged which would split the crowd in smaller groups. Information system should indicate the existence of the exact number and location of the tables at the very entrance of the dining area so that users can have ample time for taking a decision which dining area in which quadrant they should select. Separate route should be defined for leaving the used plates and washing of hands. This should be unidirectional and should not permit the crowd to once again follow the original entrance to the buffet thus adding to the chaos. Seating areas should be defined in clusters and should not allow users to drag them to a location and thus obstructing the path of movement and creating unnecessary bottleneck.

Staffs should ensure that the travel time from the kitchen to the table is minimal and with minimal criss-cross movements (Fig. 3.2). It should be ensured that crowd

Fig. 3.2 Linear movement of staffs without crossing. Photo by energepic.com: https://www.pexels.com/photo/woman-holding-tray-of-wines-2991157/

should not be allowed to accumulate, else that would create a ripple effect and can create a chaos and even stampede if the crowd pushes those in front to move faster than they can. Remember every person has a different pace for movement, and this is what creates the bottle neck when crowd from behind pushes them to move faster than they can. In a buffet this can create a mess and even lead to accidents.

3.7 Crowd Flow Management

In a gathering crowd should be managed only through ergonomic intervention, and no enforcement should be used. It's better if the layout permits crowd to disperse into small groups in well-defined zones. A separate entry, dining and exit would ensure linear movement of crowd and thus an efficient crowd flow. Separate interaction zones can be created with barricades like string and stand which gel with the ambience or flower pots with flowering plants can be aligned to create specific zones for the crowd to move, interact, dine and this would ensure an efficient crowd control. If necessary, performances could be arranged in a different zone so as to disperse a portion of the crowd. The crowd would get engaged in it, and thus there would be less of crowd flow.

3.8 Special Occasion in Dining Area

In some occasions there might be excess crowd concentration in the dining area specially when it comes to lunch times. The stereotype of users about lunch which is the meal during mid-day is very critical, because it's at this time every guest would like to have their lunch. Users perceive missing the lunch time as detrimental for their health. This could be addressed in a different way. Create multiple dining areas all over the space, allow people to sit in defined space and let food be served, or cut down on the menu list so that people finish fast! This ensures quick movement of people.

3.9 Replenishing Fast Depleting Foods

In many cases tasty food gets depleted fast and people have to wait for long, thus increasing the waiting time for the users and crowd does not move. Replenishments should be available at the table where the food is displayed so that not time is wasted in filling up the empty vessels at once. As soon as the storage reserve vessels get empty it should be immediately filed up from the kitchen. Remember attention span of users in such occasions is very low and chances of chaos is very high. Musical performance in the zone can make the waiting time less tiring as there is a attention resource from the music which can keep the crowd calm and organized. The crowd gets engaged to the music without the need for looking at it.

3.10 The Food Design Elements

Human factors in food design should focus on the look, feel and smell. The illumination of the place should be adequate to visualize the food and need not be bright. Bright vegetables and spices add color to the food and make the stand apart from everything in space. The presentation of the food is very important because it creates a context for the food and conveys a language to the guests for eliciting their hunger. Food organization adds to the wow factor in food design, and along with well-groomed staffs, illumination, music, etc. create a different ambience for enhancing the appetite of users.

3.11 Ambience Creation

Human factors in creating ambience talks about using all the senses of humans. It should look good, smell good and look clean and hygienic. The furniture element should be of wood to convey that sense of aristocracy, along with fabrics of white color which conveys the message of sterilization and cleanliness of highest extent. The illumination of the place can be a little dim but enough to see the food. This sets the mood with a light music either live or automatic which acts as a screen for hiding unnecessary noise of people talking, sound of crockeries, etc. You can arrange for open kitchen where users get a chance to see how things are prepared and allow them to participate in cooking some of their own dishes. This creates a dynamic mental model and the guests remember the experience for a long time.

3.12 Key Points

A. Human body dimensions are to be considered for designing the space and elements in space.
B. For reach consider the lower percentile value and for access the higher percentile value.
C. Human body is a mixture of different percentile value.
D. Add movements to the value that you optimize.
E. Staffs need to be physically fit and have their muscle toned to prevent injury.
F. Focus on workspace envelope of each user.
G. Perceive the space in plan and elevation.
H. Factor in human body dimensions for deciding upon the space requirement.
I. Control crowd through ergonomic design of elements in space.
J. Users need feedback as to the ingredients use and hence clarify them.
K. Food should be arranged in the sequence in which they are to be consumed.
L. Multisensory ambience can be created with a prudent mix of smell, sight, illumination and music of the place.

3.13 Practice Session

A. Ms. Smith the room service person is very short height. She finds it difficult to reach table tops, door knobs and other elements in space. How should these be designed so that everyone is able to reach them including guests?

Human Factors Direction:

1. First identify the different parts of the design of space or elements and the respective human dimensions.
2. Next refer to the respective anthropometric dimensions in the database.
3. Ask yourself if this is a question of reach? Or access?
4. Optimize accordingly, but add extra allowance for movement, clothing, etc.

B. Tourists are finding it hard to accommodate inside a tourist boat while on a cruise on the river. What would be your advice to the authorities for designing the seating system of the boat?

Human Factors Direction:

1. First identify which parts of the body are necessary for comfortable seating.
2. Refer to respective anthropometric dimensions.
3. For each part of seating decide whether its design for access or reach.
4. Optimize and add allowance for body movements.

C. Your executive chef is asking your advice as to the height of the work surface for different tasks. These are grinding (requiring force), chopping vegetables(light force) and cake decoration(precision task). What would you advise him?

Human Factors Directions:

1. Height of work station is related to the type of task done.
2. For heavy task the work surface should be near the hip.
3. For light task the work surface should be near the elbow.
4. For precision task the work surface should be close to the depression in the chest.
5. Your chef should have workstation of different height for different tasks.

Bibliography

Bentley T, Page S, Meyer D, Chalmers D, Laird I (2001) How safe is adventure tourism in New Zealand? An exploratory analysis. Appl Human Fact 32(4):327–338

Dos Santos LN, de Carvalho RJM (2012) Human factors and accessibility for people with visual impairment in hotels. Work 41(Supplement 1):1417–1424

Food and Tour

4

Overview

This chapter gives the readers an overview of the different application of human factor principles in the kitchen related to cooking and associated tasks. It discusses the issues of load handling, working in different posture and the common types of slips, trips and falls normally seen in the kitchen and simple ways of preventing them. The readers are given an insight into the human factor issues related to visiting different tourist sites. The chapter takes the readers through the challenges of group and individual travelers and the issue of guided tours. The different surprises in tours and how they are to be addressed from an ergonomic perspective are addressed here. The issues of communication human factors at tourist sites and how they should be designed for the tourists to make the place more attractive are discussed. The chapter ends with a note on the ergonomic issues for the elderly and challenged users in tourism and how they should be considered as well.

4.1 Kitchen and Dimensional Issues

Kitchen in any tour is here the food comes from. The food plays a very important role in any tour. Productivity in the kitchen is dependent upon the design of tools and workstation, people and their training, and the proper layout of the space. Apart from this the environment and how well it's handled also determine productivity in the kitchen. In the kitchen there are primarily three types of tasks performed. One is cutting and associated tasks (grinding, blending, etc.). The second is cooking followed by storage or serving. These tasks are normally sequenced as mentioned but there might be change in the sequence or back and forth movement depending upon the overall cooking demand and the type of recipe being cooked.

For tasks requiring force, the work surface should be at or near the person's hip region. For example when you grind very hard nuts like beetle nuts, fenugreek seeds, nutmeg, etc. considerable force is required, and the task is better done when the work surface is near the hip region (Fig. 4.1) which allows the user to apply more force. For the tasks which are light in nature like fine chopping of vegetables like onion, garlic, carrot, ginger, etc. it's prudent to have the work surface at the elbow level (Fig. 4.2). If the task demands precision activity for example decorating a cake with cream which requires very fine dextrous movement of the fingers of the hand, in that case the work surface has to go near the chest (Fig. 4.3).

For the purpose of cooking you have to remember that the chef has to view the base of the pan. Thus the workstation height has to factor this and consider the height of the stove and the cooking vessels so that the visual cone of the chef remains unaltered. After cooking the next task that happens is washing of utensils. This happens in the sink and thus the sink should be placed in a sequence after the cooking zone. The sink height should be such that the base of the sink is visible to the user and as washing is somewhat between light and heavy (you have to remove stubborn stains from utensils and apply force). The sink has to be at the elbow height of the user and better if it's a little below but not above.

The length of the kitchen table is decided by the number of elements to be kept and how many people would be working. If you want three people to be working side by side for chopping and cutting vegetables, then you have to take care that

Fig. 4.1 Heavy work of grinding seeds being performed little below the elbow height. Photo by Uriel Mont: https://www.pexels.com/photo/crop-faceless-man-pouring-water-into-coffee-grinder-6271318/

Fig. 4.2 Light work of vegetable chopping being performed around the elbow. Photo by Sarah Chai: https://www.pexels.com/photo/faceless-woman-chopping-bell-pepper-during-lunch-preparation-in-kitchen-7262911/

their elbows do not touch and there is gap in between. Likewise if the number of people increases then the minimum working space required is the elbow folded and upper arm extended at the side and the distance between the two elbows (Fig. 4.4).

4.2 Manual Material Handling in Kitchen

There are many types of utensils in the kitchen, and some of these are heavy and some light. One needs to be extremely careful while handling the heavy utensils. Make sure these have proper handles which can be gripped by the palm in a manner that the finger curls around the handles. There should be adequate clearance between the fingers and the body of the vessels while holding them. The person handling them should keep their back straight while moving with these heavy vessels. Any heavy weight should be dragged rather than lifted for preventing injury to the back. Care should be taken to ensure that people work with back straight and knees bent while lifting heavy weights.

Fig. 4.3 Precision work of cake decoration being performed around the chest region. Photo by Tim Douglas: https://www.pexels.com/photo/young-woman-putting-decoration-on-top-of-homemade-cake-6210907/

4.3 The Correct Human Factors of Serving Food and Beverages

While serving food on trays and carrying them over a long distance from the kitchen to the serving table, care should be taken to use all the fingers of the palm. If possible the tray could be made to rest on the palm rather than holding them with the fingers curled around the edge of the tray. This would prevent the early onset of fatigue and ensure efficient service to clients. Similarly drinks and other trays need to be held in this manner with the tray resting over a larger area of the palm, which ensure proper balance and also reduces stress on the forearm muscles.

Fig. 4.4 Elbow-to-elbow distance roughly defines one person working space requirement. Photo by Elle Hughes: https://www.pexels.com/photo/man-and-woman-wearing-black-and-white-striped-aprons-cooking-2696064/

4.4 Cleaning Activities and Postural Issues

Cleaning in the kitchen is an important task, and care should be taken not to bend the back while doing so. Mops with long handles ensure that the person maintains a straight back while cleaning the kitchen and the guest rooms. The corners and the ceiling are the places which accumulate maximum dirt and needs maximum attention. This is necessary to ensure that no dirt falls on the food being cooked. If possible mopping of the floor could be done with the legs instead of the hands for the elderly staffs, which leads to less exhaustion and fatigue. Else the mopping could be done with long handle mops which helps to maintain a straight back while working. This prevents back pain and injury. All the tasks involving repetitive activities like cutting, chopping, etc. in the kitchen demands small breaks of one to two minutes in between tasks to facilitate easy recovery from fatigue and soft tissue wear and tear.

4.5 Maintaining Tone of Muscles

Every injury related to load handling, repeated awkward postures in the kitchen can be prevented to a large extent if the muscle are toned up. Thus staffs should ensure that they regularly exercise and tone their muscles so that these toned muscle which renders stability to the joints and the back can absorb a large amount of shock and prevent injury to body parts.

4.6 Slips Trips and Fall and Their Prevention

To prevent slips, trips and fall, spillage of liquid should be prevented. The kitchen floor should be cleaned and wiped dry after every activity. Flooring should be textured with a color which ensures proper visibility of water and any liquid spilled on the floor. White-colored flooring is a better choice provided it's textured with proper illumination on the sides so as to increase visibility of any colorless fluid spilled on the floor.

4.7 Work–Rest Cycle

In spite of all changes as mentioned above in terms of design of handles of the vessels for ensuring proper grip, work surface height setting, people may still feel tired and fatigued. Many a times, people blame the bad design of the tools and workstation for this. Bus this is not correct. Our body has been designed for the purpose of hunting and gathering food. As we have now moved into a different type of lifestyle where we have to work for long hours, lift heavy loads, work in odd postures the body needs rest. Thus there should always be 2–3 min of rest in between heavy work so that the body gets a chance to recover and the person can work with increased vigor. A good night sleep and rest for the staffs in the hospitality and tourism industry is a must for preventing fatigue.

4.8 What Defines Enjoyment?

Mr. Nipen, an illiterate person, plans a short trip to the seashore with his wife and two children. They take the train from their home town and get down at the station "Breeze" which is very close to the seashore. They did not book any accommodation before as they thought that they would personally check the accommodation available and check in the cheapest one. When Nipen arrived there he was mesmerized to see the seashore. His wife asked him that they should look for an accommodation before they are gone. So Nipen along with his family goes around and finally manages a cheap accommodation within their budget. They are very happy that they have been able to make up to the seashore and also found a place to stay. They stay there for five days and thoroughly enjoyed their stay.

Mr. and Mrs. Della working professional decide to go to the seashore. They are also traveling for the first time. They book their hotels in advance to avoid any unnecessary trouble. They have a very high budget and book their room in an expensive star hotel. The couple in their middle ages take the flight to the nearest airport. From the airport they took a cab for their hotel. While they sat inside the cab they felt it was very dirty and the driver could not understand their language. As they traveled they could see the seashore, but it did not appeal to them much. They checked into the hotel and finally came to the conclusion that the hotel was substandard. Their stay at the place was for a week, but they did not like the place.

The above two incidences point to the fact that when it comes to "enjoying" a tourist place or the "tour" it's very subjective and varies with people. Enjoying a place depends on people's expectation and how far that expectation is met. In tour "comfort" of travel, and "information" to the adequate level and in a "format" that users can understand plays a very important role. You cannot "design" tourist enjoyment but you can "design "the framework which would elicit enjoyment among tourists. These could be adequate information about the place, comfortable accommodation and travel and good service. Users always give extra emphasis on the first interaction at any place. The Della couples first experience in the cab while traveling to the hotel was no good and that created a "mental model" of everything bad at the place for them. Later on they could not overcome that. Whereas for the Nipen couple who gave lot more emphasis on the seashore and its details, the other elements like travel and accommodation were trivial and thus they enjoyed their stay and the trip. They were from the beginning explorative, and for them the happiness was evoked by the endless beauty of the sea and the seashore as they had never seen them before and even did not have any specific mental model of the same.

Thus enjoyment is subjective but it's all about what people expect and what they get. This is where human factors can intervene. This can be done through providing adequate information and a preview of the experience they are about to experience. Here review of tourists and they views both positive and negative are important and helps the users in building a mental model of the place. Thus it's important that the users are given the appropriate information to the extent that is needed which is different for every person. They should also be informed about what they could expect and a mental model of the same. If they get something more they are happy, else they are not. So to your tourists make sure the features of the tourist spot, the amenities and facilities available along with what not to expect if projected before the travel, users would be less frustrated and much happier as they always draw an imaginary picture in their mind before they plan to travel.

4.9 Dealing with Group and Individual

Individual tourists want to travel on their own terms and conditions. For them every information should be provided to the utmost detail and al options should be presented to them depending upon the time available in hand. Dennis is a solo traveler. He is traveling to a reserve forest and would like to take the elephant safari. For him he needs to know apart from the features of this exciting ride any precautionary measures to be taken (anti-malarial drug for mosquito bites) as he is traveling alone. The risk associated with the safari should be indicated so that he can take a prudent decision whether or not to travel or to not travel alone but in a group. The tour operators should indicate this to remove all uncertainties among the traveler. Solo travelers also need information related to medical assistance, police and immigration if they are traveling to a different country. This feedback is key to essential customer service.

Group travelers again have a herd mentality. If a few people in the group have a weird desire or idea, then the entire group goes in that direction without thinking much. So group travelers need to be briefed before, during and after the journey the different details about the travel. At times different members of the group might demand different facilities and this has to be briefed to them before they take the tour as to what is available and what not. This prepares them for what they can and cannot expect from the tour and thus they do not make unnecessary demands.

In many cases especially at heritage sites the tour guides prefer to address a large group for explaining the different aspects of the heritage site. Here the address should first start with the exact location of the monumental site in space, celebrities and famous people associated with the site and then the history. This way the storyline and the excitement are building. For example while standing near a fort the address to the tourists could be "we are standing in front of this famous "Mango Fort" which is located in the state of Eldorado where we are now. This place is around 20 km from the sea cost and enjoys a pleasant weather all around. When the King Honest stayed here with his family he had a gala time. This fort consists of 45 rooms and I am going to take to those now…" This creates a "pattern" for the tourists, and they can relate themselves to be a part of a fort and its history. This arouses interest in them, and thus it becomes easy for the tourist guide to get the attention of the tourists very easily.

4.10 Creating the Background for Travel and Tourism

People go on tour either to unwind from their daily chores, to spend some time in peace, to experience something new. In many cases it could be all the factors. The first step toward effective tourism and travel is to sell it as a package to the tourist. For this one need to present it in a format which is in tandem with the mental model of the users. What does it mean? When users think of going on a tour what are the sequence of questions that comes? How much it would cost, what places can I see, duration of the trip, travel modality, accommodation and food, hazards if any, what would be my take away from this? What if I want to get individual attention, etc.? These questions are to be addressed in audio-visual format if possible to the tourists to lure them. If you plan to use print media then it should contain the information chunked out in bullet points supplemented with photographs of the place. This would create that interest in your clients as to why they should travel to the place you mentioned and why take your services. Talk about customization of food, talk about home food, people love carrying a piece of their home comfort as they travel! Any place is better recognized with reference to celebrities, events, or national and international status (e.g., UNESCO world heritage site).

4.11 Local Cuisine While on Tour

The local cuisine has a relationship with tours. People often remember a tour because of the cuisine itself. The cuisine because of its multisensory appeal (look and feel, taste and smell) is able to generate a better impact on users and acts as a reference point for retrieval of all information related to tourism even after a long time. You hear people saying "oh I remember when I had been to El Dora do and had that sweet dish made out of Rose Petals. The fragrance and the taste was just awesome and I would love to travel to that place once again just for that food." …So local cuisine should always be linked with tourism for a better recall value of the tourist place.

4.12 Change in Travel Plans Due to Natural Calamity

Any tour and travel have certain uncertainty associated with it. This is especially applicable for adventure tourism where the tour and travel plans have to be rescheduled, canceled or even redesigned in a different manner. Tourists can be disappointed easily with such events. Thus the best way to deal with this is to prepare the tourists in advance with what can go wrong and even if things go wrong what alternate measures are kept in place for them. Once this is made clear then users mentally get prepared and in case of any change in plans it does not affect them that much. They do not see it has a potential loss of opportunity. Ruben is on a trekking tour to mount hilarious. After the group starts trekking there is news that they should not go further and the tour is canceled. Ruben was disappointed for a transient period as he was briefed before that in case of such an event an alternate trekking route was planned. The disappointment for Rubens was much less as he was briefed and prepared for the same. So he already had that model or image in his mind that if this doesn't materialize then no potential loss is perceived and the acceptance level of the tourists are pretty high.

4.13 Information in Space About Tourist Site

Any tourist place needs proper labeling about what it is. This is needed so that anyone can get a preliminary idea about the place and get to know what he/she needs to see depending upon the time in hand. The information should be located at regular intervals as users are bad in short-term memory and tend to forget after they move. The quantum of information should be grouped (chunked) as per location. It's better if users are provided with a "you are here maps" because in that case they are able to locate themselves in the space and navigation becomes much easier. Similarly all space and elements in space that demands tourist attraction need to be labeled in very brief so as to convey what it's all about.

4.14 Elderly and Challenged Tourists

Handling of elderly and challenged tourists is a big challenge for all. Human factors can help in two ways. One is increasing accessibility to tourist places through provision for ramps; grab rails, audio input for visually impaired, textured flooring, etc. which are small changes but helps these people navigate with ease. The other one is a customized approach because ergonomic believes that every user is different and they prefer a customized treatment. This customized service is that personal touch which is very important and at times can boost the brand value of a company in the hospitality industry. I once heard a story which appealed to me. An aged couple was once traveling to the hill station on a guide tour. On arrival at the spot the elderly gentlemen suddenly fell ill. He had acute breathing problem. The guided tour was supposed to start from the next day. The couple was in stress as they did not know what to do. When the manager of the tour came to know about this he took an initiative to admit the man in the hospital and decided to send the assistant manager as a tour guide. The couple stayed back, and the manager accompanied them all through in the hospital till the man was cured and discharged from the hospital. This gesture of the manager was so well accepted by the couple that they narrated this incident on social media platforms. It ultimately leads to a wide publicity of the manager and the tour company, without them having to put any extra effort. People remember when services are rendered to them at the time of need. This one service was a memorable "landmark" in the life of the couple, and they always remembered this one tour where the man fell sick, though after that they had been to so many tourist spot. Thus personal touch of people is of immense value, and this is where "service" scores over everything else. After all humans are the most reliable!!

4.15 Different Types of Sites, Heritage, Natural, Performance and Mementos to Remember

Humans have a tendency of capturing every moment of the places they visit. They want to relive on these moments. That's the reason they take photographs and buy mementos from such places. The places to sell the mementos are always better if it's at the exit (in case of heritage sites). These mementos act as landmarks or helps in retrieval of information related to the tourist sites.

4.16 Tips for Those Who Provide Services

Tips play a very important role in tour and travel. Some want to avoid giving tips, and tour operators hesitate to ask for it fearing that it might ruin their reputation. What should be the ergonomic approach toward this? All humans are good. We need to put things in the right perspective to get the best out of people. Instead of asking for tips one could explain the amount of effort that tour operators put in for the job that they do. If you could "sell" the concept of the "importance" of tips for those who put in hard effort, a majority of the people would go for giving tips rather than refraining from it. Another option could be that before the tour disperses the tour operators could give a small performance which could be anything ranging from music, small script a joke and keeping a box in front. As people enjoy the performance the tour ends in a happy note. Tourists can become emotional and give tips in the box provided as the very placement of the box mimics that it's meant for your token of appreciation.

4.17 Key Points

A. Heavy work is to be performed at waist height, and light work to be performed at elbow height in the kitchen.
B. Adequate space in the kitchen can be calculated by the elbow to elbow distance for one person. The elbows being spread apart with the finger tip of both the hands touching each other.
C. Try using larger muscle of the body when moving heavy loads if feasible.
D. Cleaning activities should ensure that there is minimal bending and twisting of the body.
E. Exercise is the key to maintain physical fitness and keep the muscles healthy and ready for facilitating movement and even keeping the joints stable.
F. Micro breaks of few minutes in between work are effective in increasing productivity and delaying the onset of fatigue.
G. We can create the framework for enjoyment which comprises of tangible elements like information, comfort, as well as touching services.
H. Group and individual travelers need to be separately treated as travelers in group behave like the herd.
I. Tourists always want the comfort and elements of their home while they travel.
J. Give an overview to tourists before the tour starts so that they get a macro picture of the itinerary and thus they become prepared for what is to follow.
K. Challenged and elderly are to be treated as part of the normal population if possible in tours. Where needed customization could be done.
L. Mementos at tourist sites play an important role in retrieval of information related to the tourist site even years later.

4.18 Practice Session

A. A chef in the kitchen is suffering from low back pain as he is engaged in cooking all day long. The cooking platform seems to be very low compared to the chef's height. Suggest simple human factors solutions for the problem.

Directions:

1. Postural correction.
2. Check for work surface whether low.
3. Fix work surface height with reference to the task being done.
4. Suggest micro breaks between work.

B. The waiters in a restaurant are serving food to the customers. It's a busy restaurant, and one has been on their toes. As a consultant human factors specialist what human factors intervention could you suggest for increasing the waiters efficiency so as to reduce the waiting time for the customers.

Directions:

1. Increase the number of waiters.
2. Serve those having few orders first.
3. Those having long orders may to made to sit in a different area.
4. Keep serving complimentary items as. The food is prepared.
5. Give guests a feedback how long would it take for the food to arrive.

C. A group of elderly people who have just retired are planning to go on a holiday to some far off places. They are unable to decide between mountain and the sea. Your human factors expertise is solicited for helping then plan their vacation.

Directions:

1. Present to them the pros and cons of both.
2. Explain to the features of both tours.
3. Give them a cost breakdown.
4. Tell them where they can go and where they cannot.
5. Ask for their choice at the end.
6. Assure them of medical facilities and food of their choice if available.

D. A tour operator is planning a boat ride for dolphin watching on the seas. What human factors design directions would you go for to ensure a smooth and safe tour?

Directions:

1. Safety protocol to be repeated at least thrice so that its better registered.
2. Duration, cost and what to expect.
3. What can go wrong.
4. Amenities and facilities during the ride.
5. Take away from the ride.

E. At the end of a guided tour of a fort the tourist guide is hesitant to ask for tips. Suggest some human factors directions that would help in getting the tips without the need for asking and embarrassing the tourists.

Directions:

1. Keep a container with tips written in big letters.
2. Give a performance which could be musical or a magic which entertains people.
3. Thank everyone.
4. Inform what would be done with the tips, "a portion of this goes to the welfare of the homeless".

Bibliography

El-Amir J, Omar AM (2019) Investigating the relation between human factors and efficiency of hotel kitchen staff. J Faculty Tour Hotels-University of Sadat City 3(1)

Sever MM (2019) Improving ergonomic conditions at hospitality industry. Inter J Res Bus Soc Sci, 8(2):91–101 (2147–4478)

Reducing Travel Uncertainty

5

Overview

This chapter gives the readers an insight into the human factors issues related to travel and hospitality majority of which are intangible. The readers are taken through the needs and wants of tourists, their choices. The human factors issues in offering customers their choice and at the same time ensuring to provide that humane touch to the tourists are discussed in this chapter.

5.1 The Uncertainty in Travel

One of the human characteristics is they want to come out of uncertainty (Fig. 5.1). This is where adequate information at the right place with adequate feedback at regular intervals helps. Mr. Sam and Mrs. Sam are on a trip to a desert. They plan to take the desert safari on a camel. Their hotel has been booked, but the booking is on hold because it's a peak tourist season now. They have booked their train ticket to the nearest station "Sweating". The couple is traveling on their own, and once they arrive at "Sweating" they would be taken care of by a travel agency who have arranged for their accommodation, pick up from the railway station, accommodation, local sightseeing and again dropping them to the station. The problem happened because the agency told the couple that the booking is not confirmed in the hotel but asked them not to worry. The train tickets for the elderly even before ten days of the date of travel were in the waiting list but the booking officer assured them that it would be confirmed too. The couple was thus a bit jittery as they did not know what to do. They started preparing for the travel but at the same time were doing it half-hearted as they were not sure whether the travel is going to happen or not. They kept on checking their train status and repeated called up the tour agent to get an update on the booking. They were getting tensed as they never got any updates either on the status of their railway

© The Author(s), under exclusive license to Springer Nature Singapore Pte Ltd. 2024
P. Mukhopadhyay, *Human Factors in Tourism: A Layperson's Approach*, Design Science and Innovation, https://doi.org/10.1007/978-981-99-7066-7_5

booking or hotel accommodation. Just a day before the travel their train ticket was confirmed and the hotel agent called to confirm them that they can travel in peace as the hotel room is now confirmed in their name. The couple were so tensed all these days and frustrated that their joy of travel diminished and they did travel but half-hearted. After reaching the destination also though they did travel all around, did local sightseeing, etc. but did not enjoy the tour as such. So why did this happen? Possibly because there was a lack of "feedback" at regular interval. If the railway booking was presented to them in the form of how their reservation is moving toward confirmation, that would give them a peace of mind. Similarly, the booking agent could have called them every day and given feedback as to the status of the booking or a system where they could see their booking status regularly that would have helped them a lot. This reduces uncertainty associated with travel. Thus the prime objective is to reduce travel-related uncertainty. The key to becoming a successful travel and tour operator is to ensure that tourists enjoy the tour, and uncertainty is reduced through presentation of information as feedback at regular intervals. This could also be handled by meeting the client and giving them personal assurance at regular interval.

Fig. 5.1 Uncertainty of travelers decreases the pleasure of the tour. Image by **Masashi Wakui** (https://pixabay.com/users/masashiwakui-4385858/?utm_source=link-attribution&utm_medium=referral&utm_campaign=image&utm_content=2014617) from Pixabay (https://pixabay.com/?utm_source=link-attribution&utmmedium=referral&utm_campaign=image&utm_content=2014617) https://pixabay.com/photos/japan-osaka-night-asia-landmark-2014617/

5.2 The Service Angle and Human Factors in Hospitality

People are always grateful and remember any good gesture rendered to them. In tourism this "service" that is the right amount at the right time is the most important. Look into your customers' needs before they ask for it (Fig. 5.2). This is the best way to win over people. When you were young your parents fulfilled many of your wishes even before you asked for it! That has to be followed in the hospitality industry. This is where one needs to study the "users" pattern of "their needs" when they travel. For example, elderly travelers are always worried about the availability of a washroom. If this need is addressed or they are assured when the travel they find lot of trust in you and your company. Some travelers are always worried about availability of clean and hygienic food. If you can address and assure your clients this, they trust you more and more. Remember users always look for "quality" of service over quantity. Your customers are in the lookout for drinking water when they enter the hotel, provide them the facilities for the same. Your customers want a view from the room of the hotel, before they say demonstrate to them this is how it looks and ask them whether you should keep it open or closed. The next thing your customer would be interested in is dining issues, whether in room dining facilities are available if yes, the details and if not inform them. If there are any complimentary services, please brief them before you leave. Map your users' needs and address them before they ask for it.

Fig. 5.2 Providing service to customers before they ask for it. Image by: **cottonbro studio** (https://www.pexels.com/@cottonbro/) https://www.pexels.com/photo/server-pouring-hot-drink-6466292/

5.3 What Customers Aspire?

It's very important for you in the business to know what the customers aspire. Some of the things you need to keep an eye on are the indications that are evident from the body language. If tourists have traveled the whole day, then you can make out from the pace that they walk. A very sloth movement pattern indicates extreme fatigue. A person under fatigue needs rest along with some water immediately followed by beverage and some snacks. This is where you should go forward and offer your services for. This is more applicable for elderly and middle-aged traveler. For example, if a group of youngsters arrive at the hotel or are on tour, they would explore new foods at random. You need to check for their excitement level and offer them the choice of different food which an elderly or middle-aged person might not venture into because of health-related issues. Give your guests options; they will explore and enjoy. If you cannot provide certain services to your guests clearly articulate what can be provided instead of what cannot be. If asked you can reconfirm what cannot be provided. This type of positive articulation is always better accepted by any user.

5.4 Peoples Weird Choice and Demands

Humans have certain characteristics, and you need to be on the lookout for those. When people leave home and travel or stay in a different place, they carry along with them their "ways of expecting things to happen". For example, Ramesh a young man in his 20 is in a trip to the mountains. He gets up at 6am at home, and his mother gives him a glass of hot chocolate in his bed. So it has become a part of his life, and when he is on tour he expects the same and might or would demand that from the company. So, you need to be ready with people's choices and lifestyles. That's why when celebrities check in big hotels lots of background research is always done on their culture, taste of food, daily routine so that these (some of which could be extremely weird) could be addressed, and in this way you could impress your guests.

5.5 Why People Prefer Some Over the Others

There is a cut-throat competition in the tourism and hospitality industry just like any other sectors. So why would a guest pick or select a tour company, a hotel, a specific guided tour over the others. If you talk about the basic service in terms of physical comfort, destinations covered, amenities, etc. they are more or less similar. The winning edge of one company over the other needs to be looked at from the perspective of "overall experience" your guests get and its spread through word of mouth. This spreading of good will through "word of mouth" is the biggest advertiser for any company. How do you achieve that? Study your guests profile in detail. If you find that there are some with medical conditions list them down and see how you could help them. These are the people who fear to travel because of their conditions. If you find diabetics, hypertensive or wheel chair-bound people just assure them that you have made arrangements for them everywhere. For those on insulin who have to carry their insulin in icepacks are often hesitant to travel for long because of lack of facility for keeping the insulin cool. If the tour company has arrangements for this then that becomes a bonus point for them and they receive the appreciation of people and this is spread to others. Similarly, you need to be vigilant on special diets of people like low salt, sugar free, etc. and should have all information ready when tourists arrive at the tourist spot that they get what they want without having to worry about that. The assurance for these as feedback should be given to the guests before the tour starts. These unique things give your company a cutting edge over others by giving your customers a unique experience for which you create the framework in terms of your unique service to your customers. The adequacy of information in right places in the form of brochures should be provided, and the major ones like "wheel chair access" should be given in public domain.

5.6 That Something Extra for Your Customers: Customer's Sweet Spot

Mrs. Julie was traveling all alone on a sightseeing tour around the city of "Duckhawk". She suddenly saw from the bus that a person on the roadside was selling a special dessert which looked like a ball but was multicolored. She was very tempted to taste the same but realizing that the tourist bus was on the move and it cannot be stopped she just kept silent. However, the tour guide in the bus noticed this and realized Mrs. Julie's interest in the dessert. So, what he did was at the next sightseeing stoppage he arranged for a complementary dessert for all the tourists. As the announcement was sudden it was a "pleasant" surprise for not only Mrs. Julie but also for all the guests. This is called the "customers sweet spot". You study your user, try to satisfy their curiosity and desire that too packaged through a surprise which is not previously announced. People remember just "unplanned" surprises that become "landmarks" in their travel memory which they narrate to others.

5.7 The Humane Touch and Winning Hearts

In the hospitality industry the "humane" touch is the most important. How do you ensure that? It's through proper design of the space or the context followed by elements in the space. You can look the space look like a specific theme of a place. For example, tree houses, or hotels in the shape of a ship on water are interesting. This is made further interesting if you have people welcoming you at every step, guiding you to the tourist spot, explaining to you in your native language, helping you to navigate or move, etc. which are the humane touch which wins hearts

Fig. 5.3 Guests enjoying food over a music which sets the tone of the ambience. **Negative space** (https://www.pexels.com/@negativespace/) https://www.pexels.com/photo/glass-architecture-windows-lines-34112/

in this industry. The food also plays an important role specially if home-made food could be served. For example, there are some grandmother's recipe which could be served which act as a food with a different touch and thus win hearts. That childhood music which has been lost if played in the hotel lobby at night (Fig. 5.3), your guests would be nostalgic and go back to their early childhood. You might allow your guests to the kitchen and see how the food is prepared, even allow them to cook one of their dishes with assistance. This creates the framework for a unique experience which guests enjoy.

5.8 Information for Reducing Uncertainty

Make sure you provide adequate information to your guests for reducing any uncertainty. Be explicit to indicate what is free and what is to be paid. Guests are always worried about this. Information to guests/tourists as to the exact sightseeing route, prominent landmarks, hazards to be expected are to be informed before the tour starts. This creates a mental model of the tour among the tourists. Always mention the cost of entry ticket to different tourist sites so that tourists can allocate their budget accordingly. At the tourist sites also, every paid service should be prominently labeled and displayed within the visual cone. As tour operators always mention to the tourists the "approximate time taken" to see a particular sight so that tourists can take a decision whether to go or skip it.

5.9 Global Tourist

Today we live in a world where tourists travel from one corner of the world to the other. As they travel, they see the culture of other countries and also carry along with them their own culture. There ae certain things that we are taught from our childhood depending upon our geographical location (stereotype its called). For example, in some country's electrical switches in the downward direction means ON and upward direction means OFF. In other countries it's the reverse. So, when tourists from one country travel to another they might make a mistake of keeping the switch ON or OFF if not informed. This is where redundancy or "apparently useless information" may be used. For example, if we expect tourists from India (here switch in downward is On and upward is Off) to go to France its better its written on the switch ON and OFF in green and red. This bit of extra information helps the users in reducing uncertainty. Similar on roads mention the traffic rule of the country as this again differs around the world. Tell people which side of the road should they walk to their destination so that they do not meet with any accidents.

5.10 Cultural Angle

While dealing with international tourists make sure you respect their culture. In some culture certain foods are not eaten. Ensure that you never present them with such foods. There are cultures which connect a particular color with a particular attribute like "death". Do your homework and refrain from using these colors either in accommodation or in the tour. Be careful about selecting the words as some are demeaning for certain culture. So better refrain from them.

5.11 Key Points

A. Reduce travel uncertainty through proper information and feedback.
B. Render your service before its asked for.
C. Feedback and reconfirmation to users is the key when it comes to guest satisfaction.
D. Try mapping the mental model of users and address them.
E. Take utmost care to those with medical conditions while traveling and do your homework before.
F. Always walk the extra mile for clients and give them pleasant surprises.
G. The homely touch is the key to winning peoples' heart.

5.12 Exercises

A. A lone traveler from a different country arrives in a small hotel. He is not aware of the culture of the place. How would you ensure that the first interaction that he has is the best and he remembers it for the rest of his life?

Directions:

1. Know the culture and stereotype of the traveler.
2. Make sure you do not go against that as mentioned in 1.
3. Arrange for some surprises like local drink or snacks on arrival.
4. Greet them by their names.
5. If anyone is having their birthday wish them for that.

B. A group of young school children (standard 8) are on a guided tour of a city "Bakwas". This is their first tour in life on their own and without their parents. How do you ensure to incorporate that wow factor in the tour?

Directions:

1. Safety first.
2. Build in surprises in the tour, for example, magic show while traveling.
3. Arrange for their favorite food (do the homework before they arrive).
4. Take them to some special joy or fun ride of the place.
5. Give them a flavor of the local performing arts and cuisines.
6. Let them interact with one another as they travel. Facilitate this interaction further as you travel.

Bibliography

Bandy MD (1993) Reducing Accidents in the Hospitality Industry through Human factors. Hospitality & Tourism Educator 5(4):53–55

Middleton, M. (1987). Human factors systems design: An applied approach: In: Human factors in the Tourist, Agricultural and Mining Industries, Proceedings of the 22nd Annual Conference of the Human factors Society of Australia and New Zealand, Toowoomba, Queensland, December 1985, ESANZ, Carlton South, Victoria, Australia, 1985, pp 134–139. *Applied Human factors, 18*(3), 253.

Exercises in Human Factors

6

Exercise 1:

An elderly couple in their 70s is planning for a long trip to a hill station which is quite far away from their home. By train it takes two days, and by flight it takes more than two hours. As the flights are expensive the couples are planning to take the train. They are worried about booking of train tickets, hotel reservation, local sightseeing. As they are old, they are also worried about medical emergencies that if anyone is sick how is that to be handled. Thinking all these they go to a tour planner for help. Your expertise has been sought from a human factors perspective as to the approach of the tour planner in helping tourists like these so that the tour becomes enjoyable for them and hassle free.

Human Factors Directions:

The journey of the elderly couple is full of uncertainty, and they need a bit of assurance at every stage of this trip to make it enjoyable. Your first job is to see this travel and tour at a macro perspective, identify the zones which demands human factors intervention. Any travel is a long journey of uncertainty, surprises, experiences and ultimately that sense of satisfaction of getting something new and feeling refreshed. Any journey thus starts and ends at home.

So where should you look into for possible human factors intervention? Their travel is by train as it fits the budget. Can we make life easier for the elderly couple in getting them a confirmed ticket at the earliest? That solves one of their worries. The next would be to give them an overview of how you as a tour operator have planned their journey. Hand over to theme a schematic representation of day wise schedule, right from the time their train journey starts. Within the schedule include contact numbers of the person, emergency contact number. As a reconfirmatory measure share the couples phone number with the agent at the receiving end. When the journey starts the couples are anxious and at this time giving them a phone call and saying all is well and as per schedule helps them a lot. It's a two-day journey and hence a long "lag phase" which gives good opportunity for the couple to think what are the things that could go wrong. Remember they are traveling alone so give them a call thrice during the day, breakfast, lunch and dinner so as to reconfirm to them that you are keeping a track on their travel. On arrival at the destination, they need escort. This is where the travel agent should come forward to help them with their luggage and this gesture in the early morning and take them to the transport for taking them to the hotel.

At the hotel early morning check ins should be superfast as people are tired and their attention span is very low. Further details may be obtained later during the day but the tourist should be made to enter the room and rest.

On entering the room show them the amenities and facilities specially the operation of AC remote and TV remote, automatic tea/coffee maker and the number for in house dining. Later during the day present to them the schedule for the day. Users are bad in short term memory and hence do not present the complete schedule to them at one go. Present to them the schedule for that day.

While planning schedule keep in mind that these users are aged and they have a very low level of physical fitness. Thus, they fatigue very fast. Thus, journey for the sightseeing should be so arranged that they are back to the hotel for lunch, can rest for some time and then again start for further sightseeing post lunch. In this way they get less fatigue and enjoy the sightseeing. Plan the site seeing in consultation with them as too many stop overs during the journey are frustrating at this age. So, try for two to three places in a four hours trip and make sure they get adequate rest at every place, for a few minutes so that they can recover the strain of the journey to which they are normally not conditioned to at this age. You can take initiative for taking their photographs which would act as a point of reference for retrieving information about the trip of theirs when they go back.

After the tour is over, make sure you follow the same protocol that you did when they were leaving home. It's your responsibility to ensure that you keep in touch with them constantly till they enter home. After a day or two enquire about their experience, health and get feedback. This humane touch with regular feedback pays a lot of dividends in this sector.

Exercise 2:

Mr. Leto is a young professional in his mid-twenty and wishes to travel to a monumental site. He exclusively wants a tour of the monumental site along with a guide who would explain in details all facts and figures about the site. Later on Mr. Leto wants to write a book on his visit. How would you help Mr. Leto enjoy his trip?

Human Factors Directions:

The user is in an age group where they are in need of excitement, something new and is taking the trip to unwind from his daily routine. They do not need that much of hand-holding being physically fit but needs feedback, what all are available as they are ready to walk that extra mile.

As the tourist is taking the responsibility of traveling on his own till the monumental site, you should still establish contact over the phone. This is to ensure that his train has departed on time and also at least three phone calls in a day to ensure his arrival time. This provides the tourist feedback, and your "reliability "image is ingrained in his brain. Once he checks in his hotel take an appointment and meet him in person to explain to him the plans for the guided tour of the monument. This overview of what could be done helps him to take prudent decision as to how he would love to see and what extra would he opt for.

On the day of the guided trip of the monument first give him the storyline in brief with the history and significance. Then start the tour explaining to him every detail.

If tourists are left on their own to explore the site then there should be adequate information available. First the ticket counter should be labeled with the price of the ticket. Next information on the brief overview of the site close to the entrance would prepare the tourist for what they are about to experience. It's better if a rough estimate of the time taken (roughly) to see the site is given with details whether there are facilities for washroom, drinking water, cafeteria availability. Entry and exit signage should be prominent. Tourists should know in what sequence they should go and this is where ether placement of arrows at regular intervals and/or numbering of different exhibits in ascending order act as reconfirmation and helps.

Another alternative could be to use the floor for marking the direction for movement. Maps at regular intervals with a person standing give insight into the tourist's position in space and help in better navigation. You can think of different colors for specific zones like cafeteria, washrooms, emergency exit so that they pop up from the surroundings and make them prominent to the tourists.

The key to success is present information to users in steps and at regular interval. Information should be concise a few sentences preferably five to eight.

All information boards need to be mounted at the eye height of tourists and should be preferably not is justified if the quantum of text exceeds five lines to facilitate reading ease.

Exercise 3:

A group of 24 people plan for a guide tour (by luxury bus) to three different places. The group consists of young people of twenty years to elderly people of eighty. Your task is to ensure that the trip remains a memorable one (for the good reason!!) and everyone enjoys.

Human Factors Directions:

There are tangible and intangible elements involved in the tour. As the tourists would be on a guided tour on a bus you need to ensure that the dimensions of the seat are in accordance with the body dimensions. For elderly people it's necessary that they are able to recline the seats as much as possible and rest their feet. Do not give elderly the pain of climbing the bunk in case of sleeper bus. The windows should be provided with glass panes to ensure visibility for all specially those sitting near the aisles as well. Grab rails should be positioned so that people are able to hold them with their fingers curled around. For getting on the bus and getting down the grab rails are to be positioned in a manner so that tourists are able to hold them to maintain body balance. The critical part is the bed from the entrance when tourists move toward their seats from the driver's seat.

The tourists need to be briefed about the tourist place before it arrives and given a brief overview of the place. The time for which the bus would halt needs to be indicated to all and all the tourists should be requested to adjust their watches in accordance with the drivers. The bus should have a special feature on the outside in the form of a special canopy, a big balloon with the name of the company so that its exact location can be seen from a distance and tourists don't confuse the bus with that of others. This balloon can be detached and kept inside once the bus moves. Don't expect your tourists to read the bus number pate and check it for the actual bus!! We are very bad in short-term memory (like the computer RAM).

Exercise 4:

Miss Duggi from Esechiami visits Bakwas. The country she had never been before. She doesn't know the local language. Till date she had been managing to tour around with gestures. She was mainly on street food so long. Today she is hungry and wants to dine in a restaurant. She is a strict vegetarian and that's her main concern while dining. How can you help travelers like Duggi with the food and at the same time ensure she gets a good dining experience?

Human Factors Directions:

First ensure that she gets to know each and every food on the menu card. This is possible by pictorial depiction of every menu along with different ingredients which would clarify it being vegetarian or non-vegetarian. The pictorial depictions should be in a manner that they are easily understood. To further reconfirm or re-assure if the process of making the food in step-wise manner is depicted that ensures that no animal products are added.

As people do not go to restaurant for eating only, but to get an experience which is dependent on many things. These are the furniture, lighting, music, well-groomed and well-behaved people. So a light music of the place along with some paintings or photographs of the place can create an ambience and give a flavor of the place. Cleanliness is the key to the ambience, and this should be maintained at every cost. Staffs in traditional attire add to the generation of the ambience as well. Ensure that the furniture like the tables and chairs are designed in accordance with the body dimensions of the tourists to ensure maximum comfort. Guest should be welcomed at entry, and after they are finished dining they should be escorted till the main gate. This gestural way of communication is always language independent!!

Exercise 5:

Mr. and Mrs. Dado are two elderly couple who have checked into a hotel on their own. They are unable to walk properly and can somehow manage with a walking stick. They are lonely and prefer to travel alone every year and check in a budget hotel of their choice. This time they have checked in a hotel in a different country with a completely different culture. How would you ensure that they enjoy their time in the hotel?

Human Factors Directions:

In any hotel there are tangible and intangible ergonomic issues. The tangible ergonomic issue includes textured floor (to prevent slippage) and grab rails for maintaining balance all through the hotel premises. The intangible one includes concise information presentation and reconfirming the same information repeatedly.

At the entrance proper grab rails are a must for the elderly with textured flooring or carpet to increase the friction. Refrain from using pure blue as elderly are many a time pure blue color blind. Just use or add some impurities to blue. At the reception ensure that the check in is fast as elderly have very little attention span. If possible give elderly priority for checking in. The smart card access to hotel room should only include waving the card and not inserting into the slot as the later involves pinch grip which the users are not good at. The room should have bright luminaries as most of them suffer from mild-to-moderate cataracts and need more light. Bathrooms should be very well lighted, dry and with grab rails as this is the place for maximum accidents due to fall.

Meals if possible should be served in their room unless they are willing to go to the dining area. In the dining area it's prudent to serve them a sample of all different food instead of allowing them to stand in the queue along with others for food. The menu should be labeled in larger font size in upper and lower case.

Inside the room a good view of the world outside is a must, as they do not venture out much. Make sure that the newspapers and magazines reach them every day and room cleaning happens daily. They should be asked repeatedly about any service they need, and this is very important.

Exercise 6:

A group of people from Wonderland travel to Eldorado for going on a jungle safari. This is the first time they are taking the safari, and they are excited and worried as well. Your task is to ensure that the ergonomic issues are taken care of so that the safari becomes safe and enjoyable.

Human Factors Directions:

Safety is of prime importance in jungle safari. This information to be depicted at prominent places while people are waiting and should be within their visual cone. Inform tourists what they are about to see. This prepares them for the safari. Keep them informed about the rules of the jungle. If there are elderly people in the safari inform them in advance that they would not be able to use the washroom for certain time. A safari should be slow so that tourists can see, cherish and photograph. All briefings should be before the safari starts so that tourists do not ask questions while inside the jungle. Interesting information may be displayed at strategic locations in the jungle if possible like "way to jungle core", "white tigers abode", "if you couldn't see me I have seen you" and so on. When the safari ends tourists should have the options of visiting the souvenir shop which should be placed at the point where tourists get down from the vehicle after the safari. The shop should have small pieces of the jungle like animal faces on tea shirt, key ring of the jungle entrance with label, small replica of jungle safari jeeps, which are memories for tourists to cherish when they go back.

Exercise 7:

A catering service has got an order for the food and beverage of a big race meeting. The horse race is very famous, and the lunch is equally famous which is served to the guests. This is a high-volume event, and one should ensure there are no chaoses. What human factors issues should one keep in mind while arranging for the catering service for such an event?

Human Factors Directions:

For controlling the crowd the layout of the buffet table should be split up and placed at four different corners close to the wall. The sitting space should be arranged with maximum facing each other and a few single sitting facing the race course for those who travel alone. Between the dining tables there should be adequate space for two to three people to walk side by side with a little bit of overlap. All the tables should be spread along the race side with the buffet table at right angles to the dining tables. Make sure that food is served as people come to prevent overcrowding at a given time. Flowering pots may be used to create barricade near the buffet table and dining areas to separate the two and make guests form a queue. All efforts should be made to serve guests with water and beverages all through the race in their respective tables while they are watching the race.

Exercise 8:

Room service staffs in a hotel are stressed as there has been a rise in incidence of thefts by guests. When guests check out of the rooms, they carry along with them ashtray, pen stand, pillow, duvet, etc. What human factors approach would you recommend for this problem?

Human Factors Directions:

Glance at the room, top to bottom then left to right. Repeat this thrice. Then try to scan table tops and bed tops. If you detect any gap in between elements, too organized room, too untidy room, these indicate that something is wrong and possibly guests have lifted or tried to lift some elements from the room. Try entering the room thrice and scan the different surfaces, table, beds, etc. You know the pattern of each room, and thus a change in this pattern should raise the alarm.

Exercise 9:

A middle-aged professional on an office tour had checked in a hotel. He suddenly realized that he has some three hours in hand every evening. He feels like utilizing this free time for some sightseeing. How do you plan his sightseeing from an human factors perspective?

Human Factors Directions
Guest's needs are to be given the information about sightseeing based on the time taken. For example in one hour this place could be covered, and in two hours these places could be covered. This could be one layer of categorization. A further layer of categorization could be on the basis of type of tourist spot, historical, nature trail, cultural, etc. These two layers of classification help the tourist to take a decision as to what to see and when to see and thus utilizing his free time optimally for sightseeing.

Exercise 10:

A group of tourists traveling to Eldorado wants to enjoy nature and would like to opt for eco-tourism. You need to advise the tourism ministry of Eldorado as to how to help in this from an ergonomic perspective.

Human Factors Directions:

The focus should be on nature, nature trail and natural culture. It should tell the tourists the characteristics of the nature in advance so that they are mentally prepared. Parts of the nature trail could be labeled with very brief information specially related to wild animals, availability of drinking water and food and medical help. The hazards should be indicated and mentioned on the spots with specific information on the type of hazards and what is to be done in case of mishap.

Information on the spot can suggest tourists what further explorations they can go for like cycling, sailing, living in thatched houses, living with the locals, etc.

Printed in the United States
by Baker & Taylor Publisher Services